This book is an introductory textbook on the physical processes occurring in the Earth's radiation belts. The presentation is at the senior or first year graduate level, and it is appropriate for students who intend to work in some aspect of magnetospheric physics.

The treatment is quantitative and provides the mathematical basis for original work in this subject. The equations describing the motion of energetic ions and electrons in the geomagnetic field are derived from basic principles, and concepts such as magnetic field representations, guiding center motion, adiabatic invariance, and particle distribution functions are presented in a detailed and accessible manner. Relevant experimental techniques are reviewed and a summary is given of the intensity and energy spectra of the particle populations in the Earth's radiation belts.

Problem sets are included as well as appendices of tables, graphs and frequently used formulas.

Cambridge atmospheric and space science series

Introduction to Geomagnetically Trapped Radiation

Cambridge atmospheric and space science series

Editors

Alexander J. Dessler
John T. Houghton
Michael J. Rycroft

Titles in print in this series

Introduction to Geomagnetically Trapped Radiation

Martin Walt

Lockheed Missiles and Space Company and Stanford University

CAMBRIDGE
UNIVERSITY PRESS

CAMBRIDGE UNIVERSITY PRESS
Cambridge, New York, Melbourne, Madrid, Cape Town, Singapore, São Paulo

Cambridge University Press
The Edinburgh Building, Cambridge CB2 2RU, UK

Published in the United States of America by Cambridge University Press, New York

www.cambridge.org
Information on this title: www.cambridge.org/9780521431439

First published 1994
This digitally printed first paperback version 2005

A catalogue record for this publication is available from the British Library

ISBN-13 978-0-521-43143-9 hardback
ISBN-10 0-521-43143-3 hardback

ISBN-13 978-0-521-61611-9 paperback
ISBN-10 0-521-61611-5 paperback

Contents

Preface

This book is the outgrowth of teaching an introductory course in geomagnetically trapped radiation to graduate and undergraduate students at Stanford University. Because this material is well-plowed ground, the topics are presented in what is hoped to be a logical sequence, and the historical chronology is not followed. The emphasis is on the basic physical processes treated in a tutorial manner, rather than on the descriptive aspects of the magnetosphere. Thus the principles developed here should be useful regardless of further evolution in our understanding of trapped radiation. Items such as frequently used formulas and graphs of trapped particle parameters are collected in appendices for ready access and for future reference.

The rationalized MKS system of electromagnetic units is used throughout the book. Although much of the published research in this field is in Gaussian units, the author believes that the MKS system is preferable and will eventually prevail. Some exceptions to SI units are allowed, such as particle fluxes, which are expressed in $cm^{-2} s^{-1}$ rather than $m^{-2} s^{-1}$ because of convention and because detector apertures are more readily visualized in cm^2.

Inasmuch as this work is written to be a textbook rather than a review of current research, references to original works are limited. Undergraduates and first-year graduate students are more interested in understanding the material than in learning the source. The references listed are primarily those articles and books which will be useful to the student who wishes to dig deeper into any of the topics covered. The numbers listed after each of the references indicate the chapter for which that reference is particularly applicable. I apologize to my many colleagues whose work is utilized here but is not explicitly acknowledged.

Special thanks go to many of my friends who helped with various stages of the book and who provided many stimulating and informative discussions. My colleagues John Cladis, Ted Northrop, David Stern, Tim Bell and Michael Schulz were particularly helpful and patient, both in reviewing portions of the text and in working the problem sets. I must also thank the students who ferreted out inconsistencies in the lecture notes and who provided fresh perspectives on the subject.

I am grateful to the Lockheed Missiles and Space Company for authorizing my teaching efforts at Stanford University and to Stanford for providing the opportunity.

The endless number of drafts with redundant and repetitive alterations were patiently processed by my secretary, Mrs Glenda Roberts, whose competence and good nature were essential to convert rough notes into legible text.

The support and encouragement of my wife during the years involved in this book is acknowledged and appreciated.

Symbols

Symbols in bold face denote vector quantities. When only the magnitude of the quantity is used, the same symbol is used in standard type.

$\mathbf{A}$ Magnetic vector potential

A Area

$A(t)$ Asymmetric part of the time variation of the Earth's magnetic field

$\mathbf{B}$ Magnetic field

$\mathbf{B_d}$ Dipole magnetic moment

B_0 Mean value of geomagnetic field on the equator at the Earth's surface

B_{eq} Value of B in the equatorial plane not at the Earth's surface

B_m Value of B at the mirroring point of a particle

B_r, B_θ, B_ϕ Components of geomagnetic field in spherical coordinates

B_{max} Maximum value of B above the atmosphere for a given drift shell

$\mathbf{b}$ Magnetic field of a wave or perturbation on the Earth's magnetic field

c Velocity of light in vacuum

$D, D_{xx}, D_{\alpha\alpha}, D_{LL}$ Diffusion coefficients for particle transport

D_{LL}^{M} Radial diffusion coefficient produced by magnetic field perturbations

D_{LL}^{E} Radial diffusion coefficient produced by electric field perturbations

$\mathcal{D}_v$ Differential operator defined by equation (7.46)

$\mathbf{E}, \mathbf{E}_\perp, \mathbf{E}_\parallel$ Electric field intensity, perpendicular and parallel components

E_θ, E_r, E_ϕ Components of electric field in spherical polar coordinates

$E_{\phi n}$ nth Fourier coefficient in expansion of the Earth's electric field fluctuations

E Particle kinetic energy

$\hat{\mathbf{e}}_1, \hat{\mathbf{e}}_2, \hat{\mathbf{e}}_3; \hat{\mathbf{e}}_x, \hat{\mathbf{e}}_y, \hat{\mathbf{e}}_z$ and $\hat{\mathbf{e}}_r, \hat{\mathbf{e}}_\theta, \hat{\mathbf{e}}_\phi$ Unit orthogonal vectors

$\varepsilon, \bar{\varepsilon}$ Kinetic plus potential energy, average kinetic plus potential energy

$\varepsilon', \bar{\varepsilon}'$ Constant of integration equal to $\frac{1}{2}mv_\parallel^2 + \mu B$, average value

$\mathbf{F}, F_x, F_y, F_z$ Force, orthogonal components of force

$f(\mathbf{x}, t)$ Number of particles at coordinate $\mathbf{x}$ per unit $d\mathbf{x}$ at time t.

$F(\mathbf{p}, \mathbf{q}, t)$, $F(\mu, J, \phi, t)$ Distribution function of particles in phase space or adiabatic invariant space at time t.

g_n^m, h_n^m Harmonic expansion coefficients describing the core geomagnetic field

$\bar{g}_n^m$, $\bar{h}_n^m$ Harmonic expansion coefficients describing the magnetic field produced by magnetopause current systems

g_m Spatial eigenfunctions of the diffusion equation

G Geometric factor of a detector

$\mathbf{i}$ Electric current density

I Integral invariant function

$\mathbf{I}$ Electric current

$j(E, \alpha)$ Differential, directional particle flux

$j(E)$ Omnidirectional particle flux

$j(\alpha, E > E_0)$ Integral particle flux above an energy E_0

J Action integral, adiabatic invariant

$J_1 = \mu$, $J_2 = J$, $J_3 = \Phi$ First, second and third adiabatic invariants or action integrals

$\mathcal{J} = \dfrac{\partial(x_1, x_2, x_3)}{\partial(y_1, y_2, y_3)}$ Jacobian relating two sets of variables or coordinates

L Magnetic shell parameter – approximate distance from center of Earth to equatorial crossing of the field line in Earth radii)

m Mass of a particle

m_e, m_p Mass of electron, mass of proton

m_0 Rest mass of a particle

μ First adiabatic invariant, magnetic moment of a charged particle

$\mathcal{M}_E$ Magnetic moment of a dipole field

N Number density of scattering centers, number of particles

$N_1(x)$, $N_2(x)$ Geometric function of geomagnetic field defined by equations (6.28) and (6.31)

$\mathbf{n}$ Unit vector normal to surface or parallel to the radius of curvature of line

$\mathbf{p}$, $\mathbf{p}_\perp$, $\mathbf{p}_\parallel$ Particle total momentum, perpendicular and parallel components

$P_A(\Omega)$ Power spectrum of $A(t)$ evaluated at angular frequency Ω

$\mathbf{P}$ Canonical momentum of a charged particle

P_n^m Associated Legendre functions, Schmidt normalization

P_{nm} Associated Legendre functions, usual normalization

q Electric charge

q_i Position coordinate conjugate to momentum coordinate p_i

Q Source intensity for a particle distribution

r, θ, ϕ Spherical polar coordinates

$\mathbf{r}$ Position vector

R Position vector of the guiding center of a charged particle

$R(r)$, $\Theta(\theta)$, $\Phi(\phi)$ Separation functions for solution of Laplace's equation in spherical coordinates

R_c Radius of curvature of a magnetic field line

R_E Radius of the Earth

R_0 Distance from the center of the Earth to the equatorial crossing point of a magnetic field line

s Distance measured along a geomagnetic field line

s_m, s'_m Conjugate mirroring points on a magnetic field line

Δs Increment of distance along a geomagnetic field line

S Area

$S(t)$ Symmetric part of the time variation of the Earth's magnetic field

S_b Helical distance traveled by a particle during a complete bounce

t Time

$T = \dfrac{E}{m_0 c^2}$ Kinetic energy in rest mass units

$\mathbf{v}$, $\mathbf{v}_\perp$, $\mathbf{v}_\parallel$ Particle velocity, perpendicular and parallel components

v_x, v_y, v_z Orthogonal components of velocity

$\mathbf{v}_f$ Apparent velocity of magnetic field line motion

$\mathbf{v}_g$ Group velocity of a wave

$\mathbf{v}_{ph}$ Phase velocity of a wave

$V(x)$ Potential energy

$\mathbf{V}$ Velocity of moving reference frame

$\mathbf{V}_E$ Guiding center drift velocity of a particle caused by an electric field perpendicular to the magnetic field

$\mathbf{V}_G$ Gradient drift velocity

$\mathbf{V}_c$ Curvature drift velocity

$\mathbf{V}_\perp$ Guiding center drift velocity perpendicular to the magnetic field

W Relativistic total energy of a particle (kinetic energy + rest energy)

x, y, z Rectangular coordinate axes

x $\cos \alpha_{eq}$ = cosine of equatorial pitch angle of a trapped particle

$\mathbf{x}$ Generalized vector coordinate

z Atomic number of an atom

α Pitch angle of particle in a magnetic field $= \tan^{-1}(v_\perp / v_\parallel)$

α_{eq} Pitch angle of a particle measured at the equatorial plane

α_{LC} Bounce loss cone pitch angle

α_{LC}^d Drift loss cone pitch angle

β v/c

γ Relativistic factor $(1 - \beta^2)^{-1/2}$, wave growth rate

ε_0 Permitivity of free space

$\Gamma(\alpha_{eq})$	Latitude-dependent factor of radial diffusion coefficient
η	Scattering angle, longitude
η_{min}	Minimum Coulomb scattering angle determined by shielding of nuclear charge
λ	Geomagnetic latitude
λ_m	Latitude of a particle mirroring point
λ_n, λ_m	Eigenvalues of the spatial part of the diffusion equation
ν	Frequency of global magnetic or electric field fluctuations
ν_{drift}	Longitudinal drift frequency of trapped particles
μ	First adiabatic invariant $= p_\perp^2/2m_0 B$ (magnetic moment)
μ_0	Permeability of free space
$\boldsymbol{\rho}$	Gyroradius
ρ	Charge density, mass density
$\sigma(\eta)$	Cross-section for Coulomb scattering of an electron through angle η
τ_b	Bounce period of a trapped particle
τ_d	Longitudinal drift period of a trapped particle
τ_g	Gyration period of a trapped particle
ϕ	Phase angle between $\mathbf{v}_\perp$ and the magnetic field vector of a wave
Φ	Magnetic flux, third adiabatic invariant
Ψ	Probability function, scalar potential
ψ	Azimuthal angle
Ω	Solid angle
Ω, Ω_e	Angular gyration frequency, electron gyration frequency
Ω_D	Longitudinal angular drift frequency of a trapped particle
ω	Wave angular frequency
ω_d	Doppler shifted wave frequency
ω_p	Plasma frequency
ξ, ξ', ξ'', ζ	Dummy variables of integration
θ	Colatitude, polar angle
χ	$\cos \theta$ where θ is colatitude

Useful constants

e	Elementary charge	1.602×10^{-19} coulombs
m_e	Rest mass of electron	9.11×10^{-31} kg
m_p	Rest mass of proton	1.673×10^{-27} kg
c	Speed of light in vacuum	2.998×10^8 m s^{-1}
μ_0	Permeability of free space	$4\pi \times 10^{-7}$ newton s^2 coulomb^{-2}
ε_0	Permittivity of free space	8.85×10^{-12} coulomb^2 newton^{-1} meter^{-2}

Geophysical quantities

R_E	Mean radius of Earth	6.37×10^3 km
B_0	Mean magnetic field on equator at Earth's surface	3.12×10^{-5} T
$\mathcal{M}_E$	Magnetic dipole moment of the Earth	8.07×10^{22} A m²
$\dfrac{\mu_0 \mathcal{M}_E}{4\pi}$	Magnetic dipole parameter of the Earth	8.07×10^{15} T m³

Energy equivalents

$m_p c^2$ 938.3 MeV
$m_e c^2$ 0.511 MeV
1 eV 1.6×10^{-19} joules

1

The Earth's radiation belts

Introduction

The discovery in 1958 that the Earth's magnetic field contains belts of energetic ions and electrons was a major milestone in geophysics and astronomy. This discovery can be viewed as the birth of magnetospheric physics since an enormous quantity of new knowledge concerning our planet's outer environment rapidly followed the first measurements of these radiation belts. Furthermore, the interpretation of many well-known phenomena, such as the aurora, magnetic storms and ionospheric structure, required major revisions in the light of this new information. Thus, our view of planet Earth was changed and enlarged by these new discoveries, nourished by the availability of space technology.

A thin upper atmosphere and ionosphere surrounded by an empty magnetic field were no longer seen as the outer envelope of the near-Earth region. Instead, the magnetic field was found to contain a population of ions and electrons of varied origin and having a rich, dynamic behavior. The geomagnetic field in the region between the ionosphere and the solar plasma accepts charged particles from both of these sources. The electric and magnetic fields accelerate, store and transport these particles, eventually returning them to the source regions. The complexity of the processes experienced by charged particles in the geomagnetic field is truly bewildering, such that, after 30 years of sustained effort, important phenomena are still only dimly perceived. Many features that have been observed repeatedly still lack a quantitative explanation, and new discoveries will, no doubt, continue to be made.

The remainder of this introduction will be a brief, qualitative review of our knowledge of space plasmas in the region near the Earth. This overview is intended to give the reader a sense of the role of radiation belts in space science and in space technology, and to make the reader aware of the scope and variety of the phenomena to be considered. Thus, this chapter will lay the groundwork for the more detailed development of radiation belt physics presented in the body of the book.

The magnetosphere

The Earth's magnetosphere is the region of space containing magnetic fields of terrestrial origin. Electric currents in the Earth's core produce a magnetic field (about 6×10^{-5} tesla at the Earth's surface near the poles). Above the surface the geometrical pattern of the field is approximately a dipole within distances of several Earth radii from the Earth's center. The solar wind, a plasma of electrons and ions moving radially outward from the sun, impinges on the Earth's field at a velocity of $300-500$ km s^{-1}. The moving plasma compresses the field on the sunward side, flows around the magnetic barrier, and distends the field lines into a tail extending several million kilometers down-wind from the Earth. The interaction of the solar wind and the geomagnetic field is complex and the processes that control many of the characteristics are not yet understood. Nevertheless, satellite experiments conducted over the past 30 years have identified the principal features of the magnetosphere, and these are illustrated schematically in Figure 1.1.

The solar wind flow is supersonic, that is, its speed is greater than the speed of any plasma wave which can propagate in the upstream direction and warn incoming ions and electrons that an obstacle, the geomagnetic field, is about to be encountered. One might therefore expect that particles of the solar wind would impact the geomagnetic field directly and not flow more or less smoothly around the flanks. Indeed, nature solves this problem by forming a shock wave 2–3 Earth radii (R_E; $1R_E = 6.37 \times 10^3$ km is the mean radius of the solid earth) ahead of the magnetic barrier. This shock wave converts some of the directed energy of ions and electrons into thermal motion and reduces the bulk flow velocity to a value below the plasma wave speed. Thus, the plasma, after passing through the shock, is subsonic and can flow around the magnetic obstacle in much the same way that air flows around a subsonic airplane wing. The shell-like region between the shock and the magnetic barrier is called the magnetosheath. The magnetosheath is bounded on the upstream side by the shock and on the downstream side by the geomagnetic field. This inner boundary, called the magnetopause, separates the plasma and magnetic field of solar origin from the plasma and magnetic field associ-

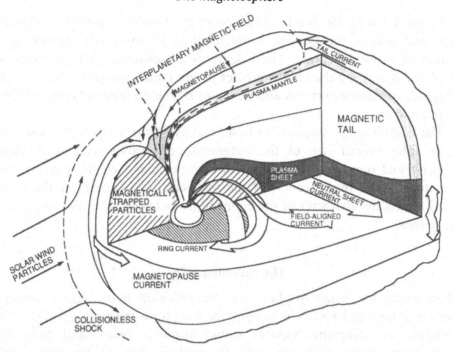

Figure 1.1. Schematic representation of Earth's magnetosphere indicating the locations and shapes of the various features.

ated with Earth. A straightforward but oversimplified calculation balancing plasma and magnetic pressures on both sides of the magnetopause predicts the location of the magnetopause in the region facing the Sun.

On the downstream side of the Earth this pressure balance approach fails. In fact, all theoretical approaches to calculating the properties of the geomagnetic tail have limitations and the picture of Figure 1.1 is largely empirical. It is clear from experiments that the geomagnetic field is drawn downstream into two great lobes, the field lines in the northern lobe pointing towards the Earth (and the Sun), while those in the southern lobe are directed away from the Earth (and the Sun). Such a magnetic configuration requires a current sheet flowing across the tail in the dawn-to-dusk direction in the equatorial plane. Other current systems must flow on the surface of the magnetosphere to separate the internal field of geomagnetic origin from the solar magnetic field carried by the solar wind. The separation of solar and Earth generated magnetic fields is not complete, however, and field lines originating in the Earth can become topologically connected to solar field lines. This process of field line merging is believed to be important for the dynamic behavior of the magnetosphere.

Figure 1.1 is an idealized illustration of the Earth's magnetic configuration and plasma regions and is intended to illustrate the location and extent of various features. These features are distinguished by magnetic field topologies, by the characteristics of the plasmas found there, by the presence of electric currents and fields, or by the presence of certain types of plasma waves.

The Earth's magnetosphere as described by Figure 1.1 is not in a steady state. The overall size of the magnetosphere varies with solar wind velocity and density, and internal instabilities cause changes in the tail structure. Also, the direction of the magnetic field carried by the solar wind affects the character of the connections between solar and terrestrial field lines.

The radiation belts

Well inside the magnetosphere lie the radiation belts, regions where energetic ions and electrons experience long-term magnetic trapping. In general, such trapping requires stable magnetic fields, and near the magnetopause the magnetic field fluctuations induced by solar wind variability prevent long-term trapping. On the low-altitude side the atmosphere limits the radiation belt particles to regions above 200–1000 km because collisions between trapped particles and atmospheric constituents slow down the trapped particles or deflect them into the denser atmosphere. Thus, in the study of trapped radiation the region of prime interest is the volume of stable magnetic field above ~200 km and below about 7 R_E at the equator. The magnetic geometry limits this volume to magnetic latitudes equatorward of about 65°. However, it is as well to remember that neighboring regions (in fact the entire magnetosphere) are involved in radiation belt phenomena. These regions supply particles for the belts and produce electric and magnetic fields that accelerate, deflect or transport the trapped particles.

The importance of trapped radiation in space science and technology

The magnetically contained ions and electrons surrounding the Earth are an integral part of the Earth's environment and play an important role in many geophysical processes. This plasma is a reservoir of energy that on occasion is released into the atmosphere producing transient aurora, airglow and ionization. The trapped ions and electrons exchange energy with plasma waves. Hence, this region of space teems with various types of waves that may be amplified, damped, refracted or reflected by the associated plasmas. The trapped particles also produce electric currents that in turn generate magnetic fields. Measurements of this variable component of the geomagnetic field have been made for decades, but many aspects were fully understood only after the discovery of trapped radiation. For example, during the main phase of a magnetic storm the depression of the magnetic field observed at the Earth's surface is caused by increased numbers of trapped ions and electrons whose motion produces a magnetic field opposing the field of the Earth's core.

In many ways, some only vaguely understood, the trapped population acts as a coupling agent transferring energy, momentum and mass between the interplanetary medium and the Earth's atmosphere. This transfer is many-faceted, sometimes occurring by the direct transport of particles, and sometimes through the intermediary of stresses in the geomagnetic field, electric currents and plasma waves. Unraveling these dynamic interactions is one of the principal immediate goals of magnetosphere research.

In spacecraft technology the energetic trapped particles have always been an important, sometimes dominant, concern. Satellite components such as solar cells, integrated circuits and sensors can be damaged by radiation, or their performance may be degraded by an increased background resulting from the passage of charged particles through the electrically active volume. A dramatic example of this vulnerability occurred in 1962 when several satellites ceased to operate after their solar cells were damaged by the enhanced radiation belt from a high-altitude nuclear explosion. At present, considerable effort is devoted to manufacturing radiation resistant electronic components for satellite equipment, primarily in order that it may survive the trapped radiation environment.

Trapped particles may also disrupt satellite operation in more subtle ways. On occasion the electrons and ions may deposit unequal charges on satellite surfaces leading to differences in the electric potential of various satellite segments. The resulting electric discharges can damage electronic components or produce spurious signals which give false instructions to spacecraft computers. These satellite 'anomalies' are often associated with unusual conditions in space such as magnetic storms.

Over the past 30 years radiation belt characteristics have been mapped extensively, and the radiation hardness of electronic components has been greatly improved. Nevertheless, the effects of the radiation belts on satellites remains a major factor in satellite lifetimes. The miniaturization of electronics and the digitization of logic circuits has made satellite instrumentation more susceptible to radiation because the energy deposited by an incoming ion may be as large as the charge representing a binary digit in the circuit. Also, the greater sophistication and efficiency of sensors has resulted in an increased sensitivity to background radiation. The Hubble Space Telescope, a major space science instrument launched in 1990, routinely has some of its sensors turned off during passage through the most intense radiation regions.

Implications for astrophysics

It is now recognized that energetic ions and electrons are a ubiquitous feature of nature, occurring wherever large-scale, fluctuating magnetic fields exist in the presence of ions and electrons. The outer planets of the solar system have radiation belts analogous to that of the Earth but with some distinctions characteristic of each planet. The Sun, whose magnetic field geometry does not support long-term, stably trapped radiation belts, nevertheless produces high-energy ions and electrons. These particles, which are accelerated by rapidly changing magnetic fields in solar active regions may reach several BeV in energy for ions and several MeV for electrons. The magnetic field configurations near active regions on the Sun probably contain these particles for several minutes, and the extended solar magnetic field throughout the solar system can hold such particles for many hours. In addition, the magnetic structures that originate on the Sun and extend through interplanetary space also accelerate ions and electrons to several MeV. Although the geometries involved in the acceleration of solar plasma and the creation of planetary radiation belts are dissimilar, some of the fundamental processes are the same. Thus, study of particle acceleration in the near-Earth region has led to improved insight into solar acceleration processes. In turn, studies of solar particles have advanced our knowledge of magnetospheric physics.

On the larger scale of galactic dimensions, energetic charged particles are also of profound importance. Best known are the cosmic rays, very high-energy particles (up to 10^{14} MeV) that permeate the galaxy. These particles can be measured directly and their energy and composition are clues to the formation, structure and evolution of the galaxy. Even when such cosmic particles cannot be detected directly, the electromagnetic radiation they produce is often observed. Thus, synchrotron radiation and X-rays from high-energy electrons reveal the presence of spinning neutron stars and even more exotic objects such as quasars and black holes. Again, while phenomena have vastly different scales than the Earth's magnetosphere, the insight obtained by the relatively advanced understanding of Earth's plasma environment has been valuable. For the foreseeable future, solar system plasmas will be the only astrophysical plasma populations in which local measurements can be made. At present, the Earth's plasma is the only region for which comprehensive, long-term data are available. The Earth is also unique in that observations have been made over most of the volume containing the energetic plasma, thereby giving a relatively complete picture of the entire structure, including the all-important boundaries. Thus a thorough understanding of the Earth's radiation belt suggests ideas and concepts which aid astronomers in the interpretation of less complete astrophysical data.

Status of radiation belt knowledge

Our present knowledge of the Earth's radiation belts has reached an advanced state, having benefited from years of research and a rapidly improving technology for performing the necessary measurements. Virtually all regions of the Earth's magnetosphere have been explored. The various populations of trapped particles have been observed, their compositions and energy spectra measured, and a long history of changes in particle distributions caused by natural variations in the Earth's magnetic and electric fields has been recorded. Many, if not all, of the physical processes have been identified. The motion of individual particles in static magnetic and electric fields is completely understood and is the subject of the first half of this book. In the second half of the book the effects of changes in the large-scale magnetic and electric fields and the resulting diffusion and acceleration of particles is considered. This topic retains some of its mystery in that the time variations of the fields have not been measured sufficiently to support a complete verification of the theoretical formulations. Also, much of the theoretical foundation rests on approximations that are not always valid.

Knowledge of the influence of various types of waves on trapped particles is in a less satisfactory state, although there is no controversy over the general principles by which waves and particles interact. Various approximations have been applied successfully when the wave amplitudes are small. However, these simplifications cannot be used for strong waves. The general case where the waves and particles exchange appreciable energy has not been fully investigated within magnetospheric geometry and is an active area of current research.

In general, our knowledge of radiation belt processes decreases with

increasing distance from the Earth. The inner radiation belt region, which
was the first region to be explored by spacecraft and which is the most
stable of the trapping regions, is rather well mapped and explained. Much
of the well-established theory presented in this volume finds its application
in the near-Earth radiation regions. As one moves outward through the
magnetosphere, the belts become subject to larger variations and exhibit a
wider variety of behavior. The regions connected by magnetic field lines to
higher latitudes participate in the dumping of particles to sustain the
aurora and are supplied with energetic ions drawn from the atmosphere.
At the outer boundary of the radiation belts, time variations are large, and
the magnetic field is greatly distorted by the solar wind pressure. The
radiation belts' populations are fed by solar plasma and depleted by the
escape of trapped particles into the magnetosheath. In this outer region
much remains to be accomplished, both experimentally and theoretically.

   It is generally correct to say that the exploratory phase of radiation belt
research is ending, and we are now entering an era of detailed investiga-
tion in which theory will be confronted by more comprehensive and
precise data. The data most needed now are simultaneous observations
made from key locations throughout the magnetosphere. Such measurements will involve multispacecraft observations as well as multi-instrument
measurement of particle fluxes, electric and magnetic fields, and the
characteristics of waves over a broad frequency range. Only when such
data are available and have been explained can one be confident that all
the important processes in radiation belt physics have been identified.

Problems

1. Assume that the solar wind has a number density of 5×10^6 ions and electrons
   m^{-3} and a velocity of $400 \, \mathrm{km \, s^{-1}}$. This moving plasma applies a dynamic
   pressure of $2\rho v^2$ to the magnetopause, where ρ is the mass density and v is the
   solar wind speed. This pressure is balanced by a magnetic field pressure of
   $B^2/2\mu_0$ inside the magnetopause at the subsolar point. What is the value of B
   at that location?

2. About half the magnetic field just inside the magnetopause originates in the
   Earth's core and half is produced by currents along the magnetopause. If the
   core field at the subsolar point is given by $B_{eq} = 3 \times 10^{-5}(R_E/r)^3 T$, where R_E
   is the radius of the Earth and r is the distance from the Earth's center to the
   point in question, what will be the distance from the center of the Earth to
   the subsolar magnetopause?

3. The most energetic cosmic ray particles have energies of about 10^{14} MeV. This energy is equivalent to that of a 10^{-2} kg marble dropped from what height?

4. The solar constant, the amount of radiant energy striking a 1 m^2 surface at the Earth is 1.39×10^3 J m^{-2} s^{-1}. Assuming that all this radiation is absorbed by the atmosphere, land or oceans, what is the total radiant power (in J s^{-1}) which the Sun can supply to the Earth. Using the data from problem 1 and assuming the frontal part of the Earth's magnetosphere is a half sphere 15 R$_E$ in radius, what is the total power of the solar wind striking the magnetosphere?

2

Charged particle motion in magnetic and electric fields

Introduction

The fundamental equation describing the motion of a charged particle in magnetic and electric fields is the Lorentz equation

$$\mathbf{F} = \frac{d\mathbf{p}}{dt} = q(\mathbf{v} \times \mathbf{B} + \mathbf{E}) \qquad (2.1)$$

For rationalized MKS units the force $\mathbf{F}$ is in newtons, the charge q is in coulombs, the electric field $\mathbf{E}$ is in volts m^{-1}, the velocity $\mathbf{v}$ is in $m\,s^{-1}$ and the magnetic field $\mathbf{B}$ is in webers m^{-2} or tesla. A list of symbols used throughout this book is given in the list of symbols (pp. xiii–xvi).

For some simple field geometries, equation (2.1) can be integrated directly to give the trajectory of the particle. However, for the geomagnetic field such an integration is not possible, and one must resort to approximations. Fortunately, for radiation belt particles whose energy is so low that the magnetic field appears almost uniform, an efficient approximate theory has been developed. The results of this theory will be presented in stages in the following chapters. First, the motion of a charged particle in simplified magnetic and electric fields will be considered. This discussion will illuminate the fundamental reasons for particle trapping. It will be seen that in general the particle executes a rapid circular motion while at the same time the center of the circle moves through the electric and magnetic fields. Equations for the motion of this so-called 'guiding center' give a quantitative description of the motion of the guiding center and confirm the trapping properties of the geomagnetic field. In Chapter 4 the adiabatic invariance approximation is introduced. This theory describes the long-term trajectory of the guiding center, although it does not give the guiding center velocity or indicate where the guiding center will be at a given time.

In the first part of this chapter we will use equation (2.1) to obtain the particle motion in fields with simple geometries. The extension of this motion to the geomagnetic field will then be easy to understand. Equation (2.1) can be separated into components parallel and perpendicular to the magnetic field giving

$$\left(\frac{d\mathbf{p}}{dt}\right)_{\parallel} = q\mathbf{E}_{\parallel} \tag{2.2}$$

and

$$\left(\frac{d\mathbf{p}}{dt}\right)_{\perp} = q(\mathbf{v} \times \mathbf{B} + \mathbf{E}_{\perp}) \tag{2.3}$$

Uniform magnetic field

Assume **B** is uniform and constant and that **E** is zero. For these conditions equations (2.2) and (2.3) become

$$\left(\frac{d\mathbf{p}}{dt}\right)_{\parallel} = 0 \tag{2.2'}$$

$$\left(\frac{d\mathbf{p}}{dt}\right)_{\perp} = q(\mathbf{v} \times \mathbf{B}) = q(\mathbf{v}_{\perp} \times \mathbf{B}) \tag{2.3'}$$

Integrating (2.2') gives

$$\mathbf{p}_{\parallel} = \text{constant}$$

indicating that the particle moves parallel to **B** at a constant speed. The momentum change in equation (2.3') is perpendicular to $\mathbf{v}_{\perp}$. Therefore, $\mathbf{v}_{\perp}$ is constant in magnitude, and the trajectory is a circle of radius ρ when projected on to a plane perpendicular to **B**. The centrifugal force must balance the Lorentz force giving

$$\frac{mv_{\perp}^2}{\rho} = qv_{\perp}B$$

or

$$\rho = \frac{p_{\perp}}{Bq} \tag{2.4}$$

The radius ρ is an important parameter characterizing particle motion. It is frequently called the gyroradius or cyclotron radius. The angular frequency of the gyration motion, the gyrofrequency, is

$$\Omega = 2\pi\frac{v_{\perp}}{2\pi\rho} = \frac{Bq}{m} \text{ radians s}^{-1} \tag{2.5}$$

Note that in the non-relativistic case ($m = \text{constant}$), Ω is independent of particle energy. Thus, in a uniform magnetic field with no electric field the

particle describes a helix, the circular motion in the plane perpendicular to
B being superimposed on a uniform motion parallel to **B**. The pitch angle
of the helix is the angle between the particle velocity and the magnetic
field and is given by $\alpha = \tan^{-1}(v_\perp/v_\parallel)$. Particles with large pitch angles
near 90° move essentially in circles. If the pitch angle is near 0°, the helix is
more open and the particle motion is predominently parallel to **B**.

The helical motion described above is the primary motion of trapped
particles in the geomagnetic field because the non-uniformities in the field
are small over distances the length of the gyroradius. However, even weak
gradients in the geomagnetic field introduce deviations in the particle
motion, and these deviations lead to particle trapping.

Uniform magnetic and electric fields

If $\mathbf{E}_\parallel$ is constant, the parallel equation (2.2) leads to uniform acceleration
along a field line

$$\mathbf{p}_\parallel(t) = \mathbf{p}_\parallel(t = 0) + q\mathbf{E}_\parallel t \tag{2.6}$$

Such parallel fields are rarely found in the trapping region of the
magnetosphere, although they are important in accelerating particles in
the aurora.

Moderate electric fields perpendicular to a uniform **B** result in a drift
motion perpendicular to both **B** and **E** (Figure 2.1). This effect can be

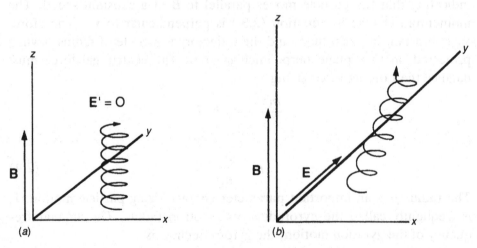

Figure 2.1. Motion of a charged particle in perpendicular electric and magnetic
fields. (*a*) Particle motion as observed in a frame of reference moving in the *x*
direction with velocity $\mathbf{V}_E = \mathbf{E} \times \mathbf{B} / B^2$ such that $\mathbf{E}' = 0$ in the moving frame. (*b*)
Particle motion as observed in a stationary frame of reference in which an electric
field **E** is present.

understood most easily by using a Lorentz coordinate transformation to eliminate the electric field (see Appendix A).

Let $\mathbf{B} = B\hat{\mathbf{e}}_z$, $\mathbf{E} = E\hat{\mathbf{e}}_y$. If the primed quantities denote values measured in a reference frame moving at some arbitrary velocity $\mathbf{V}$ perpendicular to $\mathbf{B}$, then the electric field in the moving system is

$$\mathbf{E}' = \mathbf{E} + \mathbf{V} \times \mathbf{B} \tag{2.7}$$

To eliminate the electric field in the moving frame, $\mathbf{V}$ is chosen such that $\mathbf{E}' = 0$. The vector product of (2.7) and $\mathbf{B}$ (setting $\mathbf{E}' = 0$) gives

$$0 = \mathbf{B} \times \mathbf{E} + \mathbf{B} \times (\mathbf{V} \times \mathbf{B})$$
$$= \mathbf{B} \times \mathbf{E} + (\mathbf{B} \cdot \mathbf{B})\mathbf{V} - (\mathbf{B} \cdot \mathbf{V})\mathbf{B}$$

Because $\mathbf{B} \cdot \mathbf{V} = 0$, the required frame velocity is

$$\mathbf{V} = \frac{\mathbf{E} \times \mathbf{B}}{B^2} \equiv \mathbf{V}_E \tag{2.8}$$

In a frame moving at velocity $\mathbf{V}_E$ the electric field vanishes and the particle executes the helical motion described earlier. In a stationary frame the motion is a deformed gyromotion drifting at velocity $\mathbf{V}_E$ in the x direction. The reason for the drift can be traced to a distortion of the circular gyromotion by the electric field.

In its gyromotion a positive particle has greatest energy and largest gyroradius when it is at maximum excursion in the y direction (Figure 2.2). Viewed in the $x-y$ plane the trajectory accumulates displacement in

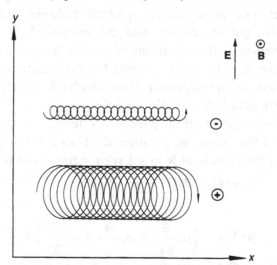

Figure 2.2. Explanation of $\mathbf{E} \times \mathbf{B}/B^2$ drift mechanism. Radius of curvature increases as particle kinetic energy is increased.

the x direction. A negative particle circles in the opposite sense and has its largest gyroradius while at minimum y, thus drifting in the positive x direction also. As is apparent from equation (2.8), all charged particles drift in the $\mathbf{E} \times \mathbf{B}$ direction with a velocity depending only on $\mathbf{E}$ and $\mathbf{B}$ and independent of particle charge, mass or velocity. Note also that the drift is perpendicular to $\mathbf{E}$ so that, on average, no energy is gained or lost during the drift.

Equation (2.8) is valid as long as $|\mathbf{V_E}|/c \ll 1$. If the electric field is so large that $|\mathbf{V_E}|/c$ is appreciable, relativistic equations must be used to calculate the particle motion and the description used here does not apply. In the Earth's magnetosphere electric fields are never so large that (2.8) cannot be used.

Inhomogeneous magnetic field

The most interesting effects from the standpoint of trapping occur when $\mathbf{B}$ is not uniform. Even for electrons and protons of many Mev energy, $\rho \ll R_E$, and the geomagnetic field experienced by the particle during a gyration is almost uniform. Nevertheless the slight deviations from helical motion which are produced by ∇B accumulate over time and lead to important perturbations in the otherwise helical motion of the particle.

One is generally not interested in the individual gyrations of the particle but wishes to follow its path over an extended trajectory very much larger than the gyroradius. This motivation leads to the concept of a 'guiding center' in which one separates the particle behavior into the circular motion about the 'guiding center' and the motion of the guiding center itself. The derivation of the equations of motion for the guiding center is sketched here for $\mathbf{E} = 0$, $\partial \mathbf{B}/\partial t = 0$ and for non-relativistic particles, as this case will illustrate the approximations involved. For the more general case and for more details see Northrop, 1963.

Express the position $\mathbf{r}$ of a particle in terms of its instantaneous gyroradius ρ and the center of gyration $\mathbf{R}$. Thus $\mathbf{r} = \mathbf{R} + \rho$. Expand the magnetic field in the vicinity of $\mathbf{R}$ in a Taylor series about $\mathbf{R}$

$$\mathbf{B}(\mathbf{r}) = \mathbf{B}(\mathbf{R}) + \rho \cdot \nabla\mathbf{B}(\mathbf{R}) + \ldots \qquad (2.9)$$

where

$$\rho \cdot \nabla\mathbf{B} = \left(\rho_x \frac{\partial}{\partial x} + \rho_y \frac{\partial}{\partial y} + \rho_z \frac{\partial}{\partial z}\right)\mathbf{B}$$

Substitute (2.9) into (2.1), with $\mathbf{E} = 0$ and denote the time derivatives by dots above the quantity.

$$m(\ddot{\mathbf{R}} + \ddot{\boldsymbol{\rho}}) = q(\dot{\mathbf{R}} + \dot{\boldsymbol{\rho}}) \times [\mathbf{B}(\mathbf{R}) + \boldsymbol{\rho} \cdot \nabla \mathbf{B}(\mathbf{R}) + \ldots] \qquad (2.10)$$

The basic assumption that $\rho(|\nabla \mathbf{B}|/B) \ll 1$ allows one to neglect the higher-order terms in the Taylor expansion. Let $\hat{\mathbf{e}}_1$ be a unit vector in the direction of the magnetic field at $\mathbf{R}$; the unit vectors $\hat{\mathbf{e}}_2$ and $\hat{\mathbf{e}}_3$ then form an orthogonal coordinate system such that $\hat{\mathbf{e}}_1 \times \hat{\mathbf{e}}_2 = \hat{\mathbf{e}}_3$ (see Figure 2.3). The gyroradius $\boldsymbol{\rho}$ will be in the $\hat{\mathbf{e}}_2$–$\hat{\mathbf{e}}_3$ plane and can be expressed as

$$\boldsymbol{\rho} = \rho(\hat{\mathbf{e}}_2 \sin \Omega t + \hat{\mathbf{e}}_3 \cos \Omega t) \qquad (2.11)$$

Repeated differentiations with respect to time give

$$\dot{\boldsymbol{\rho}} = \Omega\rho(\hat{\mathbf{e}}_2 \cos \Omega t - \hat{\mathbf{e}}_3 \sin \Omega t) + \sin \Omega t \frac{\mathrm{d}}{\mathrm{d}t}(\rho\hat{\mathbf{e}}_2) + \cos \Omega t \frac{\mathrm{d}}{\mathrm{d}t}(\rho\hat{\mathbf{e}}_3) \quad (2.12)$$

$$\ddot{\boldsymbol{\rho}} = \Omega^2\rho(-\hat{\mathbf{e}}_2 \sin \Omega t - \hat{\mathbf{e}}_3 \cos \Omega t) + \dot{\Omega}\rho(\hat{\mathbf{e}}_2 \cos \Omega t - \hat{\mathbf{e}}_3 \sin \Omega t)$$

$$+ 2\Omega \cos \Omega t \frac{\mathrm{d}}{\mathrm{d}t}(\rho\hat{\mathbf{e}}_2) - 2\Omega \sin \Omega t \frac{\mathrm{d}}{\mathrm{d}t}(\rho\hat{\mathbf{e}}_3) + \sin \Omega t \frac{\mathrm{d}^2}{\mathrm{d}t^2}(\rho\hat{\mathbf{e}}_2)$$

$$+ \cos \Omega t \frac{\mathrm{d}^2}{\mathrm{d}t^2}(\rho\hat{\mathbf{e}}_3) \qquad (2.13)$$

Equations (2.11), (2.12) and (2.13) for $\boldsymbol{\rho}$, $\dot{\boldsymbol{\rho}}$ and $\ddot{\boldsymbol{\rho}}$ are now substituted into equation (2.10) and the resulting equation is averaged over time, integrating over a complete cyclotron period with t going from 0 to $2\pi/\Omega$. Because

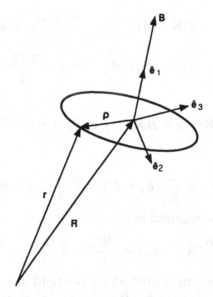

Figure 2.3. Diagram defining vector coordinate system for particle gyration in an inhomogeneous magnetic field.

all terms in $\boldsymbol{\rho}$, $\dot{\boldsymbol{\rho}}$ and $\ddot{\boldsymbol{\rho}}$ contain either $\sin \Omega t$ or $\cos \Omega t$ as factors, the averages of these quantities are zero:

$$\langle \boldsymbol{\rho} \rangle = \langle \dot{\boldsymbol{\rho}} \rangle = \langle \ddot{\boldsymbol{\rho}} \rangle = 0$$

After time averaging, equation (2.10) becomes

$$m\ddot{\mathbf{R}} = q[\dot{\mathbf{R}} \times \mathbf{B}(\mathbf{R})] + q\frac{\rho^2 \Omega}{2}[\hat{\mathbf{e}}_2 \times (\hat{\mathbf{e}}_3 \cdot \nabla)\mathbf{B} - \hat{\mathbf{e}}_3 \times (\hat{\mathbf{e}}_2 \cdot \nabla)\mathbf{B}] + \ldots$$

$$(2.14)$$

Additional, somewhat tedious, vector algebra reduces this expression to

$$m\ddot{\mathbf{R}} = q[\dot{\mathbf{R}} \times \mathbf{B}(\mathbf{R})] - q\frac{\rho^2 \Omega}{2}\nabla B + \ldots \tag{2.15}$$

where B is the magnitude of the magnetic field.

Equation (2.15) is the basic equation of motion for the guiding center. The higher-order terms which have been neglected are generally not important for radiation belt studies, and these additional terms will not be indicated in subsequent equations. However, it is as well to recognize that the equations derived here and on the pages immediately following contain approximations which become less valid as the gyration radius increases. The more useful forms of equation (2.15) are obtained by separating the equation into perpendicular and parallel components. The perpendicular component is obtained by taking the vector product of (2.15) with $\hat{\mathbf{e}}_1$:

$$m\ddot{\mathbf{R}} \times \hat{\mathbf{e}}_1 = q(\dot{\mathbf{R}} \times \mathbf{B}) \times \hat{\mathbf{e}}_1 - \frac{q\rho^2 \Omega}{2}\nabla B \times \hat{\mathbf{e}}_1$$

$$= q\{(\hat{\mathbf{e}}_1 \cdot \dot{\mathbf{R}})\hat{\mathbf{e}}_1 B - B\dot{\mathbf{R}}\} - \frac{q\rho^2 \Omega}{2}\nabla B \times \hat{\mathbf{e}}_1$$

or

$$Bq\{\dot{\mathbf{R}} - (\hat{\mathbf{e}}_1 \cdot \dot{\mathbf{R}})\hat{\mathbf{e}}_1\} = Bq\dot{\mathbf{R}}_\perp = m(\hat{\mathbf{e}}_1 \times \ddot{\mathbf{R}}) + \frac{q\rho^2 \Omega}{2}\hat{\mathbf{e}}_1 \times \nabla B$$

Hence

$$\dot{\mathbf{R}}_\perp = \frac{m}{Bq}(\hat{\mathbf{e}}_1 \times \ddot{\mathbf{R}}) + \frac{\rho^2 \Omega}{2B}\hat{\mathbf{e}}_1 \times \nabla B \tag{2.16}$$

To the approximation required here,

$$\ddot{\mathbf{R}} = \frac{d}{dt}(\dot{\mathbf{R}}_\perp + \dot{\mathbf{R}}_\parallel) \approx \frac{d\dot{\mathbf{R}}_\parallel}{dt} = \hat{\mathbf{e}}_1\frac{dv_\parallel}{dt} + v_\parallel^2\frac{\partial \hat{\mathbf{e}}_1}{\partial s} \tag{2.17}$$

where s is the distance measured along the field line, which need not be straight. With this expression for $\ddot{\mathbf{R}}$ inserted into (2.16) the perpendicular velocity then becomes

$$\dot{\mathbf{R}}_\perp = \hat{\mathbf{e}}_1 \times \left(\frac{\rho^2 \Omega}{2B} \nabla B + \frac{m}{Bq} v_\parallel^2 \frac{\partial \hat{\mathbf{e}}_1}{\partial s} \right)$$

$$= \hat{\mathbf{e}}_1 \times \left(\frac{mv_\perp^2}{2qB^2} \nabla B + \frac{m}{Bq} v_\parallel^2 \frac{\partial \hat{\mathbf{e}}_1}{\partial s} \right) \qquad (2.18)$$

For obvious reasons the first term in (2.18) is called the gradient drift and the second term the curvature drift. A more transparent interpretation of these quantities will be given shortly.

The parallel component of equation (2.15) is extracted by forming the scalar product with $\hat{\mathbf{e}}_1$.

$$m\ddot{\mathbf{R}} \cdot \hat{\mathbf{e}}_1 = -\frac{q\rho^2 \Omega}{2} (\nabla B) \cdot \hat{\mathbf{e}}_1$$

or

$$\frac{dv_\parallel}{dt} = -\frac{1}{2} \frac{v_\perp^2}{B} (\nabla B)_\parallel \qquad (2.19)$$

Equation (2.19) shows that for motion parallel to **B** the guiding center of a particle is accelerated in a direction opposite to the gradient of the magnetic field. If the particle is moving into a stronger field, it will be repelled, regardless of the sign of the particle's charge or the direction of the magnetic field.

Equations (2.8), (2.18) and (2.19) give the guiding center drifts of primary interest to radiation belt physics. As mentioned before, they contain approximations which may become important as the particle energy and gyration radius increases. In particular, the equation for parallel motion (2.19) is less exact than the equation for guiding center motion perpendicular to the magnetic field (2.18). Whenever these equations are used together when numerically tracking a particle trajectory, it is necessary to use a more accurate version of (2.19).

Additional terms neglected in equation (2.17) may be important if there are large electric fields or if the magnetic field changes direction with time.

Simple, physical interpretations for the gradient and curvature drifts of equation (2.18) and the 'mirroring' forces in (2.19) can be given. These drifts are analogous to the electric field drift in that the gyroradius ρ varies during the circular motion. This effect can be seen in Figure 2.4 where ∇B is in the y direction and **B** is in the $-\hat{\mathbf{e}}_x \times \hat{\mathbf{e}}_y$ direction. The trajectory of a positive particle illustrates how the smaller gyroradius at larger y (and larger B) leads to a drift in the $\mathbf{B} \times \nabla B$ direction. The magnitude of the drift can be estimated directly as follows. Because the trajectory is symmetric about a line parallel to the y axis and passing through the point

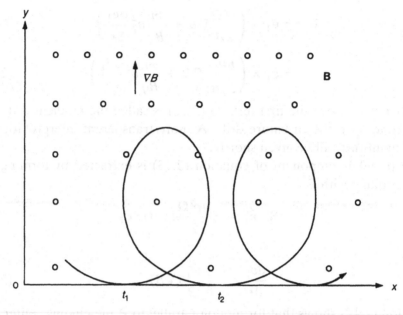

Figure 2.4. Drift motion perpendicular to **B** and to ∇B in an inhomogeneous magnetic field.

where the trajectory crosses itself, there is no net motion in the y direction. Hence, averaging the force over a gyration period should result in no net force in the y direction. The times t_1 and t_2 denote minimum y positions (taken as $y = 0$) and therefore the start and end of a cycle. If F_y is the force in the y direction,

$$\int_{t_1}^{t_2} F_y \, dt = 0 = \int_{t_1}^{t_2} q \frac{dx}{dt} B(y) \, dt \tag{2.20}$$

Because $B(y)$ does not change appreciably in a gyroradius, $B(y)$ can be approximated by the first two terms of a Taylor series:

$$B(y) = B_0 + y \frac{\partial B}{\partial y} \tag{2.21}$$

where B_0 is the value of the field at $y = 0$ and $\partial B/\partial y$ is a constant. Equation (2.21) is substituted into equation (2.20), giving

$$0 = q \int_{t_1}^{t_2} \frac{dx}{dt} B_0 \, dt + q \int_{t_1}^{t_2} \frac{dx}{dt} y \frac{\partial B}{\partial y} \, dt$$

Therefore,

$$\int_{x(t_1)}^{x(t_2)} dx = -\frac{1}{B_0} \frac{\partial B}{\partial y} \int_{x(t_1)}^{x(t_2)} y \, dx \tag{2.22}$$

The right-hand integral is the negative of the area enclosed by the curve, and if the drift in a gyroperiod is small, this area is equal to $\pi\rho^2$. The distance traveled in the x-direction during one gyration is therefore

$$\Delta x = \frac{1}{B_0}\frac{\partial B}{\partial y}\pi\rho^2$$

and the time Δt required for this cycle is $2\pi\rho/v_\perp$. Therefore, with appropriate substitutions and setting $B_0 = B$ the gradient drift term is

$$\mathbf{V}_G = \frac{\Delta\mathbf{x}}{\Delta t} = \frac{mv_\perp^2}{2qB^3}(\mathbf{B}\times\nabla B) \qquad (2.23)$$

in agreement with the first term in equation (2.18).

Note that the gradient drift term is in a direction perpendicular to $\mathbf{B}$ and to ∇B. Hence, this drift will carry particles along a line of constant B. This characteristic will be useful later in tracing the drift paths of particles near the Earth's equatorial plane. In contrast to the electric field drift, the gradient drift velocity depends on the particle energy and charge. In the non-relativistic case the gradient drift velocity is proportional to the perpendicular energy. Negative particles and positive particles drift in opposite directions. The drifts therefore produce electric currents, even in neutral plasmas.

The curvature drift term (the second term in equation (2.18)) depends on the magnetic field changing direction with distance s along the field line. A heuristic derivation of this term follows from the assumptions that the guiding center 'almost' follows a field line and the field line curvature therefore exerts a centrifugal force on the particle. The force is perpendicular to $\mathbf{B}$ and lies in the plane of curvature. The geometry is given in Figure 2.5, where $\mathbf{n}$ is a unit vector in the direction of the radius of curvature. The guiding center motion parallel to $\mathbf{B}$ exerts a centrifugal force

$$\mathbf{F} = \frac{mv_\parallel^2}{R_c}\mathbf{n} \qquad (2.24)$$

where R_c is the radius of curvature of the field line. The force on the particle is equivalent to that from an electric field of magnitude $\mathbf{E}_c = mv_\parallel^2\mathbf{n}/qR_c$. Such an electric field results in a drift velocity

$$\mathbf{V}_c = \frac{\mathbf{E}_c\times\mathbf{B}}{B^2} = \frac{mv_\parallel^2}{qR_c}\cdot\frac{\mathbf{n}\times\mathbf{B}}{B^2} \qquad (2.25)$$

This result is the same as the last term in (2.18) because

$$\frac{\partial\hat{\mathbf{e}}_1}{\partial s} = \frac{\partial\hat{\mathbf{e}}_1}{\partial\theta}\cdot\frac{\partial\theta}{\partial s} = -\mathbf{n}\left(\frac{1}{R_c}\right)$$

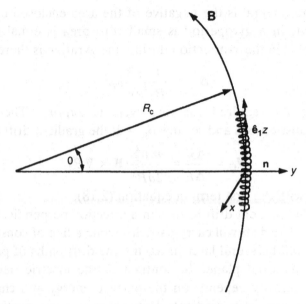

Figure 2.5. Geometry for curvature drift calculation. Drift is perpendicular to **B** and to the field line radius of curvature R_c.

If the region of space under consideration does not contain electric currents, a more convenient expression for $-\mathbf{n}/R_c$ can be derived. With the geometry shown in Figure 2.5 and utilizing $\partial B_z/\partial y = \partial B_y/\partial z$ obtained from $\nabla \times \mathbf{B} = 0$ (only valid if $\mathbf{J} = 0$ and $\partial \mathbf{E}/\partial t = 0$),

$$\nabla_\perp B = \mathbf{n}\frac{\partial B_z}{\partial y} = \mathbf{n}\frac{\partial B_y}{\partial z} = -\mathbf{n}\frac{B}{R_c} \qquad (2.26)$$

The curvature drift term in (2.18) thus reduces to

$$\mathbf{V}_c = \frac{mv_\parallel^2}{qB^3}(\mathbf{B} \times \nabla B) \quad \text{if } \nabla \times \mathbf{B} = 0 \qquad (2.27)$$

Note the similarity between equations (2.23) and (2.27) in that both drifts are in the same direction and have the same dependence on B and q. They differ, however, in their pitch-angle dependence. Particles with large pitch angles respond primarily to the gradient drift, while the curvature term is more important for particles with large $v_\parallel$.

In the parallel motion equation ((2.19)) the effect of the gradient parallel to **B** also has a simple interpretation. With the geometry of Figure 2.6 the magnetic field is in the z direction with a gradient in the $-z$ direction. A particle executing a circle about the z axis will experience a small component of **B** parallel to its gyroradius. When the particle crosses

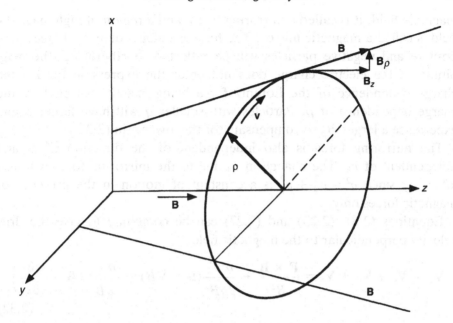

Figure 2.6. Mirror effect produced by a converging magnetic field. Gyrating particle senses a magnetic field component in the ρ-direction which deflects the particle away from the ∇B direction.

the $y = 0$ plane, the component of **B** in the ρ or x direction is

$$B_\rho = \rho \frac{\partial B_x}{\partial x} \qquad (2.28)$$

At $x = 0$

$$B_\rho = \rho \frac{\partial B_y}{\partial y} \qquad (2.29)$$

Since B_ρ must be constant around the particle orbit

$$B_\rho = \frac{\rho}{2} \left[\frac{\partial B_x}{\partial x} + \frac{\partial B_y}{\partial y} \right] = -\frac{\rho}{2} \frac{\partial B_z}{\partial z} \qquad (2.30)$$

where use is made of the Maxwell equation $\nabla \cdot \mathbf{B} = 0$.

The force in the z direction will be given by

$$\mathbf{F}_z = q(\mathbf{v} \times \mathbf{B}_\rho) = -q v_\perp \frac{\rho}{2} \frac{\partial B_z}{\partial z} \hat{\mathbf{e}}_z$$

$$= -\frac{m v_\perp^2}{2B} \frac{\partial B}{\partial z} \hat{\mathbf{e}}_z \qquad (2.31)$$

and since $\partial B/\partial z < 0$ is negative, the force is in the positive z direction. Because the force tends to reflect a particle out of a region with high magnetic field, it is called a mirroring force, and a region of high magnetic field is called a magnetic mirror. The force is independent of charge; both positive and negative particles will be reflected. Furthermore, the magnitude of the electric charge does not enter the expression for $\mathbf{F}_z$, the charge dependence of the Lorentz force being exactly canceled by the charge dependence of ρ. Particles with smaller q will have larger ρ and experience a larger B_ρ to compensate for the lower q in (2.31).

The mirroring force is also independent of the direction of B and independent of $v_\parallel$. The change in $v_\parallel$ due to the mirroring force will also affect $v_\perp$ since $v^2 = v_\perp^2 + v_\parallel^2$ is a constant of motion in the presence of magnetic forces only.

Equations (2.8), (2.23) and (2.27) can be combined to give the drift velocity perpendicular to the magnetic field.

$$\mathbf{V}_\perp = \mathbf{V}_E + \mathbf{V}_G + \mathbf{V}_c = \frac{\mathbf{E} \times \mathbf{B}}{B^2} + \frac{mv_\perp^2}{2qB^3}(\mathbf{B} \times \nabla B) + \frac{mv_\perp^2}{qB^2}\left(\mathbf{B} \times \frac{\partial \hat{\mathbf{e}}_1}{\partial s}\right)$$

(2.32)

When $\nabla \times \mathbf{B} = 0$,

$$\mathbf{V}_\perp = \frac{\mathbf{E} \times \mathbf{B}}{B^2} + \frac{m}{qB^3}\left(\frac{v_\perp^2}{2} + v_\parallel^2\right)\mathbf{B} \times \nabla B$$

(2.33)

The drift and mirror equations derived in this chapter are the essential elements which lead to particle trapping in the Earth's magnetic field. Although other effects are important, such as time variations of the electric and magnetic fields, the three magnetic effects of gradient drift, curvature drift and mirroring are the primary controlling factors leading to long-term trapping. The electric field drift term, which applies equally to all particles, is of most interest for low-energy particles, for which the magnetic drift terms are smaller. Because the electric field drift is in a direction perpendicular to $\mathbf{E}$, the particle will move along an equipotential surface and thus conserve energy. However, if magnetic curvature or gradient drifts are also present these forces will in general carry the particle across electric equipotentials and alter the particle energy.

The drift terms derived here allow one to understand geomagnetic trapping. The scale size of the magnetosphere is so large compared to the gyroradii of trapped particles that the magnetic field experienced by the particle during a gyration is almost uniform. Thus, an energetic particle introduced into the geomagnetic field circles about the field direction while moving parallel to the field line. The parallel motion will take the particle towards the poles of the Earth, where the increased magnetic field

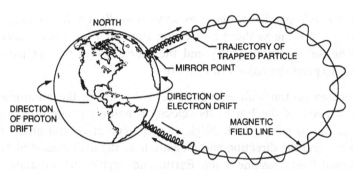

Figure 2.7. Trajectory of trapped electrons and protons experiencing magnetic mirroring and gradient and curvature drifts in the geomagnetic field.

intensity causes the particle to be reflected. The bounce motion between mirrors in the polar regions is superimposed on the much slower curvature and gradient drifts, which are perpendicular to the magnetic field, and for $(\nabla \times \mathbf{B} = 0)$ are perpendicular to the gradient of B in the plane perpendicular to the magnetic field. If the Earth's field were symmetric about the polar axis, these drifts would be entirely in the longitudinal direction. However, distortion in the geomagnetic field alters this simple result, and the drifts will have components in the latitude and altitude directions. For the Earth, the gradient and curvature drifts are eastward for electrons and westward for protons. The overall motion is sketched in Figure 2.7. Note, however, that for clarity the size of the gyroradius is greatly exaggerated in this diagram.

Problems

1. An electron of rest mass m_0 moving at velocity v has a total energy $W = \gamma m_0 c^2$, where $\gamma = 1/\sqrt{(1 - \beta^2)}$ and $\beta = v/c$. If T is the kinetic energy in rest mass units, show that

$$\beta = \frac{\sqrt{(T(T + 2))}}{T + 1}$$

Find the velocity of a 5.1 keV electron, a 51 keV electron, a 510 keV electron and a 94 MeV proton.

2. The guiding center expression for the gradient drift velocity is valid only if $\rho|\nabla B|/B \ll 1$. Assuming that the Earth's magnetic field in the equatorial plane is $B(r) = 3 \times 10^{-5}(R_E/r)^3$ tesla, where $R_E = 6.37 \times 10^3$ km, find the value of r at which $\rho|\nabla B|/B = 0.1$ for a 94 MeV proton moving $\perp$ to $\mathbf{B}$ in the equatorial plane.

3. A magnetic field parallel to the z axis varies as $B(z) = B_0 \exp(az)$. A proton of velocity v moving in the $+z$ direction starts at $t = 0$ with a pitch angle of 45°. Using the equation for $F_\parallel$ find the time it takes for the particle to be reflected. Express the answer in terms of v.

4. A low-energy proton with energy of $0.1\,\mathrm{eV}$ is above the atmosphere in the equatorial plane of the Earth. Its velocity is entirely in the direction perpendicular to the magnetic field. Neglecting the curvature and gradient B magnetic drifts, find the direction and magnitude of the drift caused by the Earth's gravitational field. Assume a flat Earth, and neglect the variation of gravity with altitude. The geomagnetic field points towards the north and has a magnitude of 3×10^{-5} tesla.

5. A particle of mass m and charge q is at rest in a uniform magnetic field **B**. At $t = 0$ a uniform electric field perpendicular to **B** is switched on. Transform to a moving coordinate system to remove **E** and describe the motion of the particle in the moving coordinate system. By transforming back to the original frame show that the maximum energy that the particle can acquire is $2m(E/B)^2$.

3

The geomagnetic field

Earth's magnetic field is produced by a number of current systems. By far the most important from the standpoint of trapped radiation is the interior current system of the Earth's dynamo. Deep in the Earth the convection of hot, conducting material forms a system of moving conductors. The motion of these conductors across the geomagnetic field induces electric currents, which in turn reinforce the magnetic field. Thus, convection driven by heat acts as a self-exciting dynamo to produce the main part of the geomagnetic field. While this field is steady on a time scale of less than a year, secular changes do occur and have been measured directly for several centuries. Systematic variations in the shape of the field are taking place and the overall geomagnetic field is becoming weaker at a rate which, if continued, will cause the Earth's field to disappear in about 2000 years. However, the present downward trend may only be a temporary fluctuation and could change at any time.

There is clear geological evidence that the polarity of the geomagnetic field has reversed at irregular intervals of about one million years. The reason for these reversals is not known, but explanations proposed include internally driven oscillations similar to those causing the solar sunspot cycles. Another mechanism suggested invokes meteoritic bombardment which would disrupt the existing polarity and allow a new polarity to develop in a random direction.

Most planets have magnetic fields. With the exception of Venus, Mars and possibly Pluto, all planets have strong fields. The Sun also has an overall magnetic field in addition to the more localized magnetic patches near sunspots and active regions. Thus, one finds that large, rotating bodies with conducting liquid or gaseous interiors generally have magnetic fields produced by internal current systems.

While a theory capable of predicting planetary fields has not been

25

forthcoming, a curious relationship between planetary magnetic fields and planetary parameters has been noted. Known as the magnetic Bode's law, it proposes a linear relationship between the logarithm of the magnetic moment and the logarithm of the angular momentum of a planet. Like the original Bode's law relating planetary orbits, this relationship is entirely empirical. There is, of course, some reason to expect that the larger the planet and the more rapid the rotation, the larger will be the internally generated magnetic field.

Other sources of the geomagnetic field are crystal rocks, which retain permanent magnetism, and currents in the ionosphere and magneto-sphere. The magnetism of geological deposits is generally weak and localized and has little influence on the geomagnetic field at high altitude. However, rock magnetism is quite important as a tracer of the Earth's magnetic history and as a clue to mineral deposits. The ionospheric current systems are strongest in the polar regions. They, too, have little influence on trapped radiation, and their average values can be incorporated into the model of the core field. The other major current systems of the magnetosphere are the ring current, the magnetopause current, the tail current sheet and the field aligned currents connecting the polar ionospheres to the magnetosphere. The ring current system consists of an extended band of trapped particles circling the Earth in the magnetic equatorial plane between $3R_E$ and $5R_E$ from the Earth's center. Since particles drifting around the Earth produce a magnetic field which inside their drift orbits opposes the Earth's field, the ring current manifests itself as a decrease in the magnetic field observed on Earth. From the ground the most noticeable evidence of the ring current is the decrease in the geomagnetic field observed on Earth during magnetic storms, when the number of trapped particles in the ring current region increases. The magnetopause current system flows along the boundary between the solar plasma and the Earth's magnetic field. It is largely responsible for the overall shape of the magnetosphere boundary. Similarly, the tail current sheet, which is an east-to-west current flowing in the equatorial plane and extending from about $10R_E$ to some large distance down the tail, separates the northern and southern magnetic tail lobes. The field aligned current systems flow along magnetic field lines and connect magnetospheric plasma with the ionosphere. These currents are important in transferring energy and momentum between the magnetosphere and ionosphere and are confined to field lines entering the atmosphere at high latitudes.

Computer programs have been developed to calculate the geomagnetic field at arbitrary positions. These magnetospheric models include many or all of the above mentioned current systems and will be discussed in the following section.

Representation of the Earth's interior field

In magnetospheric research it is essential to have a convenient method of calculating values of the geomagnetic field at arbitrary locations. For proving theorems and evaluating concepts, a simple, analytic representation of the field is usually adequate. However, when more accurate and detailed knowledge is required, rather extensive numerical algorithms are used to generate field values. In this chapter some of the customary ways of describing the Earth's magnetic field will be presented.

The frequently used multipole expansion was first utilized by Gauss and is obtained as follows. One wishes to express the magnetic field by a concise formula containing many adjustable parameters, but the field must satisfy Maxwell's equations regardless of the values of the parameters.

Maxwell's equations for magnetic fields are

$$\nabla \cdot \mathbf{B} = 0 \tag{3.1}$$

$$\nabla \times \mathbf{B} = \mu_0 \varepsilon_0 \frac{\partial \mathbf{E}}{\partial t} + \mu_0 \mathbf{i} \tag{3.2}$$

In the steady state $\partial/\partial t = 0$, and if there are no currents passing through the magnetosphere in the region of interest, (3.2) becomes

$$\nabla \times \mathbf{B} = 0 \tag{3.3}$$

If the curl of a quantity is zero, that quantity can be represented as the gradient of a scalar. Therefore, for some potential field $\psi(\mathbf{r})$ equation (3.3) will hold if

$$\mathbf{B}(\mathbf{r}) = -\nabla \psi(\mathbf{r}) \tag{3.4}$$

From equation (3.1),

$$\nabla \cdot \mathbf{B} = -\nabla^2 \psi = 0 \tag{3.5}$$

A choice of ψ which satisfies equation (3.5) will automatically satisfy equations (3.1) and (3.3). The solution to equation (3.5) (Laplace's equation) in spherical coordinates is found by assuming ψ to be separable into a product of functions of the three coordinates

$$\psi = R(r)\,\Theta(\theta)\,\Phi(\phi) \tag{3.6}$$

Inserting this expression into (3.5), dividing by ψ and separating variables leads to the various functions

$$R(r) = Ar^n + \frac{B}{r^{n+1}} \tag{3.7}$$

$$\Phi(\phi) = C \cos m\phi + D \sin m\phi \tag{3.8}$$

$$\Theta(\theta) = E\, P_n^m(\cos\theta) + F\, Q_n^m(\cos\theta) \tag{3.9}$$

where P_n^m and Q_n^m are Legendre functions. The quantities A, B, C, D, E and F are constants of integration and n and m are separation constants. The separation constants are not arbitrary: m must be an integer if $\Phi(\phi)$ is to be single valued. The Legendre functions in equation (3.9) must have integer n or they will diverge. Similarly, $A = F = 0$ when representing the Earth's field since the functions they multiply are infinite at $r = \infty$ or at $\theta = 0°$ and $180°$. Thus, the general form of (3.6) appropriate to representation of the core field is

$$\psi = \sum_{n=1}^{\infty} \sum_{m=0}^{n} \frac{1}{r^{n+1}}\, P_n^m(\cos\theta)\,(C_n^m \sin m\phi + D_n^m \cos m\phi) \tag{3.10}$$

The $n = 0$ term is excluded to avoid divergence of **B** at the origin.

In geomagnetism it is cusomary to write (3.10) as

$$\psi = R_E \sum_{n=1}^{\infty} \left(\frac{R_E}{r}\right)^{n+1} \sum_{m=0}^{n} (g_n^m \cos m\phi + h_n^m \sin m\phi)\, P_n^m(\cos\theta) \tag{3.11}$$

where the Legendre functions have the Schmidt normalization, namely

$$P_n^m = \left[\frac{(n-m)!(2 - \delta_{0,m})}{(n+m)!}\right]^{1/2} P_{n,m}$$

where $P_{n,m}$ are the normal Legendre functions and $\delta_{0,m}$ is unity for $m = 0$ and zero otherwise. The constant factor R_E is included in equation (3.11) to give g_n^m and h_n^m the dimensions of a magnetic field. The coefficients g_n^m and h_n^m are adjusted to fit experimental values of the magnetic field sampled on a worldwide basis. Although the sum over n extends to infinity, the magnitude of the terms drops rapidly with increasing n. The first few coefficients for a reference field in 1985 in nT (10^{-9} tesla) are

n	m	g_n^m	h_n^m
1	0	−29 877	
1	1	−1903	5497
2	0	−2073	
2	1	3045	−2191
5	−	~200	~150

Most magnetic models include ≥ 48 terms. The models also give the secular variations for the terms, namely the values of dg_n^m/dt and dh_n^m/dt, so field calculations can be done for any time epoch.

Because of the $r^{-(n+1)}$ dependence of ψ the importance of the higher-order terms decreases rapidly with distance from the Earth. Hence, much of trapped radiation theory is developed based on the dominant $n = 1$ or dipole term. However, when comparing radiation belt measurements taken at various points, it is necessary to use a magnetic field representation which is accurate enough to specify the geomagnetic field. Computer programs exist for calculating **B** at any point in space and for tracing geomagnetic field lines. The most recent programs contain additional functions to represent the various current systems in space as well as those in the interior of the Earth. These programs usually specify current distributions or potential functions for the various current systems and add the magnetic fields produced by these systems to give an overall magneto-spheric field.

The dipole field

The lowest order, but dominant, term in (3.11) is the dipole term with $n = 1$, $m = 0$. Because many of the important features of the radiation belts can be illustrated with a dipole field, some useful relations for this field will be derived. The dipole potential from (3.11) is

$$\psi = R_E \left(\frac{R_E}{r} \right)^2 g_1^0 \cos \theta \tag{3.12}$$

where the distance r is measured from the center of the dipole and θ is the polar angle or colatitude (see Figure 3.1). The magnetic field **B** is equal to $-\nabla \psi$. In spherical polar coordinates the components of **B** are

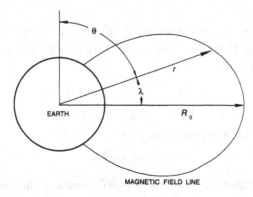

Figure 3.1. Dipole field coordinate system.

$$B_r = -\frac{\partial \psi}{\partial r} = 2 \left(\frac{R_E}{r}\right)^3 g_1^0 \cos \theta = -2B_0 \left(\frac{R_E}{r}\right)^3 \cos \theta \qquad (3.13)$$

$$B_\theta = -\frac{1}{r}\frac{\partial \psi}{\partial \theta} = \left(\frac{R_E}{r}\right)^3 g_1^0 \sin \theta = -B_0 \left(\frac{R_E}{r}\right)^3 \sin \theta \qquad (3.14)$$

where B_0 is the mean value of the field on the equator at the Earth's surface. The components are negative since the direction of the field is in the minus $\hat{e}_\theta$ direction and in the northern hemisphere it is in the minus $\hat{e}_I$ direction. When considering only the magnitude of the field the minus signs can be omitted. For the Earth $B_0 = 3.12 \times 10^{-5}$ T. The dipole field is symmetric about its axis so that $B_\phi = 0$ everywhere.

The strength of a dipole can be characterized by the magnetic moment $\mathcal{M}$ whose units are ampere meters^2. In terms of the dipole moment the radial field component is

$$B_r = -\left(\frac{4\pi}{\mu_0}\mathcal{M}\right)\frac{2\cos \theta}{r^3}$$

Some authors designate the quantity $(4\pi/\mu_0)\mathcal{M}$ as the dipole moment so some confusion exists in published reports. The use of the equatorial surface field B_0 in (3.13) and subsequent equations avoids this ambiguity.

The intensity of the dipole field at any point in space is

$$B = \sqrt{(B_r^2 + B_\theta^2)} = B_0 \left(\frac{R_E}{r}\right)^3 \sqrt{(1 + 3\cos^2 \theta)} \qquad (3.15)$$

The field intensity falls as r^{-3} with distance above the Earth and at constant r the intensity increases as one moves towards the poles. For a given value of r the field strength is twice as high over the poles as it is over the equator. Note that in equation (3.15) the r and θ dependences are separable. Hence, along any constant latitude line the field decreases as r^{-3}.

The equation for a geomagnetic field line in spherical coordinates is obtained by noting that the ratio of the lengths of the $\hat{e}_r$ and $\hat{e}_\theta$ components of the field line is

$$\frac{dr}{rd\theta} = \frac{B_r}{B_\theta} = \frac{2\cos \theta}{\sin \theta} \qquad (3.16)$$

This equation can be integrated to give

$$r = R_0 \sin^2 \theta \qquad (3.17)$$

where R_0 is the value of r when $\theta = 90°$, namely the distance from the dipole to the point where the field line crosses the equatorial plane.

Expressed in terms of latitude λ,

$$r = R_0 \cos^2 \lambda \tag{3.18}$$

The distance along a field line for a dipole can be obtained analytically by integrating the equation for a distance element ds

$$ds = \sqrt{((dr)^2 + (r\,d\theta)^2)}$$

where dr and $r\,d\theta$ are constrained to be on a field line. Expressing dr in terms of $d\theta$ by differentiating the field line equation (3.17):

$$dr = 2R_0 \sin\theta \cos\theta\,d\theta$$
$$ds = \sqrt{(4\,R_0{}^2 \sin^2\theta \cos^2\theta + R_0{}^2 \sin^4\theta)}\,d\theta$$
$$= R_0\sqrt{(1 + 3\cos^2\theta)} \sin\theta\,d\theta \tag{3.19}$$

By changing the variable to $\chi = \cos\theta$ equation (3.19) can be integrated from $\chi = 0$ (equatorial plane) to some off-equator value χ giving

$$s = \int_0^\chi ds = \frac{R_0}{2}\left[\chi\sqrt{(1 + 3\chi^2)} + \frac{1}{\sqrt{3}}\ln\left(\sqrt{(1 + 3\chi^2)} + \sqrt{3}\,\chi\right)\right] \tag{3.20}$$

The intesity of the magnetic field along a field line passing through R_0 is obtained as a function of colatitude by inserting r from equation (3.17) into (3.15), giving

$$B(\theta) = B_0\left(\frac{R_E}{R_0}\right)^3 \frac{\sqrt{(1 + 3\cos^2\theta)}}{\sin^6\theta} \tag{3.21}$$

$$= B_{eq}\frac{\sqrt{(1 + 3\cos^2\theta)}}{\sin^6\theta}$$

$$= B_{eq}\frac{\sqrt{(1 + 3\sin^2\lambda)}}{\cos^6\lambda} \tag{3.22}$$

B_{eq} is the value of B in the equatorial plane at distance R_0 from the dipole. From these equations it is apparent that the magnetic field along a field line increases monotonically with latitude as one moves from the equator to the poles.

If the Earth's field were a pure dipole located at the center of the Earth, contours of constant B on the Earth's surface would be lines of constant latitude. However, asymmetries in the interior current system introduce higher-order terms and the actual isointensity lines are as shown in Figure 3.2. Much of the distortion is caused by the fact that the magnetic axis is not aligned with the spin axis of the Earth and the center of the magnetic dipole is not at the center of the Earth. The poles are over northern Canada and southern Australia on the Mercator projection. Note, particularly, the large region of reduced field on the east coast of South America.

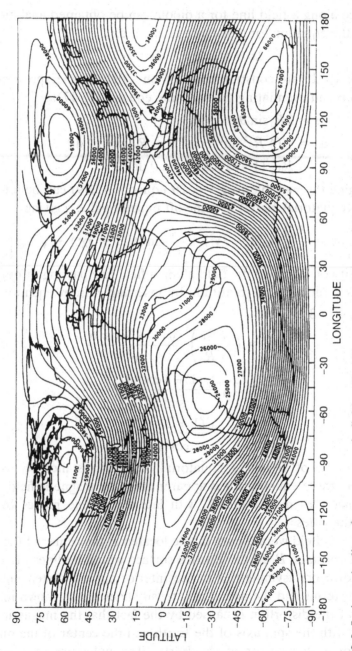

Figure 3.2. Isointensity lines for geomagnetic field at Earth's surface. Magnetic field intensity is in units of 10^{-9} tesla. In a pure centered dipole field the isointensity lines would be horizontal.

This anomaly has important consequences for the structure of the radiation belts at low altitude.

Representation of the external current systems

Although the electrical currents flowing inside the Earth are the most important currents in the production of the geomagnetic field, the external currents flowing in the magnetosphere also influence the field. These external currents are important at distances beyond about 4R and become dominant near the magnetopause or deep in the tail region. As stated earlier, the primary magnetospheric current systems are the magnetopause currents, the ring current, the tail current sheet and the field aligned currents. Although trapped particles rely on the properties of the core field for containment, the distortion of the core field by the external current systems is also relevant, and it is frequently necessary to include these systems in models of the geomagnetic field. Also, because the short-term time variations of **B** in the trapping region are caused by changes in these external currents, modeling the field fluctuations requires modeling the external field produced by these current systems.

Because the magnetopause currents are confined to the surface of the magnetosphere, the magnetic field produced by these currents can be expressed as the gradient of a scalar function for positions within the magnetosphere. The solution of Laplace's equation is again obtained by a spherical harmonic expansion as in equation (3.11). However, in this case positive powers of the radial distance are needed which give a potential field of

$$\psi_{\text{ex}} = R_E \sum_{n=1}^{\infty} \left(\frac{r}{R_E}\right)^n \sum_{m=0}^{n} (\bar{g}_n^m \cos m\phi + \bar{h}_n^m \sin m\phi) \, P_n^m(\cos \theta) \quad (3.23)$$

With appropriate constants, this potential, when added to the core potential, represents a confined magnetosphere.

The ring current, the tail current and the field aligned current systems cannot be described by scalar potentials as they flow within the region of space under consideration and, therefore, $\nabla \times \mathbf{B}$ does not vanish. However, various analytical models have been developed to describe these magnetic fields and to allow rapid computer generation of field quantities. These analytic expressions contain parameters which are adjusted to fit magnetic field measurements. The most sophisticated models allow one to compute the external field for various conditions of geomagnetic activity, solar wind characteristics and angle between the dipole axis and the solar wind flow direction.

Problems

1. Assuming that the Earth has a centered dipole field consider the magnetic field line which passes through the equatorial plane at $2R_E$ (R_E = radius of Earth). Find

   (a) The latitude at which the field line reaches the Earth.
   (b) The latitude where $B(\lambda) = 2B_{eq}$ (B_{eq} = value of B on the field line at the equator). (Don't try to give an analytic solution. Use trial and error to obtain an approximate answer.)
   (c) The distance along the field line from the equator to the surface of the Earth.
   (d) The distance along the field line from the equator to the center of the Earth.

2. An auroral physicist wishes to aim her ground based camera upward parallel to the geomagnetic field. Fortunately, she lives on a planet where the magnetic field is a pure dipole located at the center of the planet and aligned with the geographic axis. If she works at a latitude of 67°, at what zenith angle (south of the local vertical) must she point her camera?

3. A 1 MeV proton with $v_\parallel = 0$ is in the equatorial plane at $r = 2R_E$ and drifts around the Earth in a time interval τ_d. Find the energy of a proton at $3R_E$, $v_\parallel = 0$ which will drift around the Earth in the same time interval.

4. The dipole field of the Earth can be expressed by the formulas for the two components:

$$B_r = -2B_0\left(\frac{R_E}{r}\right)^3 \cos\theta\, \hat{e}_I$$

$$B_\theta = -B_0\left(\frac{R_E}{r}\right)^3 \sin\theta\, \hat{e}_\theta$$

where θ is the polar angle or *co-latitude*; B_0 is the magnetic field intensity on

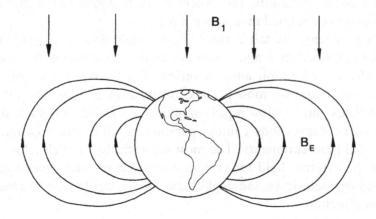

the Earth's surface at the equator. A uniform magnetic field of intensity B_1 and direction downward over the north magnetic pole is added to the Earth's field:

(a) What is the formula for B_r in the combined fields?
(b) What is the formula for B_θ in the combined fields?
(c) What is the formula for the total field strength?

4

Adiabatic invariants

Introduction

The guiding center equations of motion developed in Chapter 3 are an enormous improvement over the Lorentz force equation for describing the long-term behavior of particles in inhomogeneous magnetic and electric fields. When applied to the Earth's field, they clarify the cause of magnetic trapping. However, the drift and mirroring force equations do not allow long-range prediction of particle location, particularly in fields without axial symmetry. For example, without numerically integrating the guiding center equations over many bounces and over many degrees of longitudinal drift, a procedure likely to introduce errors, it is not possible to predict where a particle launched on a field line over Africa will be when it has drifted over the United States.

Missing in the theory described thus far are 'constants of motion' analogous to the conservation of energy, momentum and angular momentum in mechanical systems. Adiabatic invariants fill the role of the required constants of motion, and their use is essential in research on trapped radiation.

Fortunately, in mechanical systems undergoing periodic motion in which the forces change very slightly over a period, approximate constants do exist. These are called 'adiabatic invariants', implying that their values are constant provided the forces directing the motion are altered infinitely slowly. The concept of an adiabatic invariant is illustrated in the following heuristic example of an frictionless particle confined in a potential well whose shape undergoes slow variations with time. Let the one-dimensional potential well be described by curve 1 in Figure 4.1, where $V(x)$ is the potential energy as a function of the spatial variable x. A particle with total energy $\varepsilon = V(a) = V(b)$ will then oscillate between turning points a

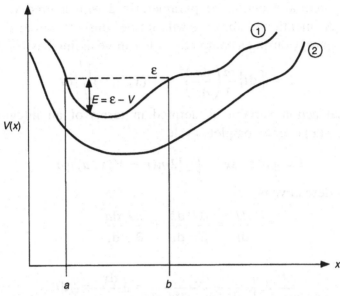

Figure 4.1. Frictionless particle confined in a potential well in which the shape of the potential changes slowly from curve 1 to curve 2.

and b, its kinetic energy at any point in its periodic motion being $E = \varepsilon - V(x)$.

The equation of motion of the particle is

$$m\frac{\mathrm{d}^2x}{\mathrm{d}t^2} = -\frac{\mathrm{d}V}{\mathrm{d}x} \qquad (4.1)$$

The total energy ε and velocity are

$$\left.\begin{aligned} \varepsilon &= \tfrac{1}{2}m\left(\frac{\mathrm{d}x}{\mathrm{d}t}\right)^2 + V(x) \\ v(x) &= \sqrt{(2E/m)} \end{aligned}\right\} \qquad (4.2)$$

Suppose, now, that the shape of the potential well changes slowly with time, the change being very small during a single period of the particle's motion. However, the cumulative change can be large, eventually altering the shape of the well to curve 2 in Figure 4.1. Since the moving walls of the well can either add or subtract energy from the particle, the total particle energy, kinetic plus potential, is not a constant. The issue is, then, what will the energy of the particle be when the shape of the well is at curve 2? Are there other quantities which remain almost constant during the variations in $V(x)$? It will be seen that the classical action integrals fill this role.

Let the potential function be parameterized with a time variation as $V(x, a(t))$. As mentioned above, ε will not be constant, but at any time t, the average energy during a complete cycle can be defined as

$$\bar{\varepsilon} = \oint dt \left[\frac{m}{2} \left(\frac{dx}{dt} \right)^2 + V(x, a(t)) \right] \bigg/ \oint dt \tag{4.3}$$

The classical action variable is defined in terms of an integral of the momentum $p(x)$ over a complete cycle

$$J = \oint p(x)\, dx = \oint \sqrt{[2m(\bar{\varepsilon} - V(x, a))]}\, dx \tag{4.4}$$

and its time derivative is

$$\frac{dJ}{dt} = \frac{\partial J}{\partial \bar{\varepsilon}} \frac{d\bar{\varepsilon}}{dt} + \frac{\partial J}{\partial a} \frac{da}{dt} \tag{4.5}$$

But

$$\frac{\partial J}{\partial \bar{\varepsilon}} = \oint \frac{dx}{\sqrt{[2(\bar{\varepsilon} - V)/m]}} = \oint \frac{dx}{v} = \oint dt \tag{4.6}$$

and

$$\frac{d\bar{\varepsilon}}{dt} = \oint dt \left[\left(m \frac{d^2x}{dt^2} + \frac{\partial V}{\partial x} \right) \frac{dx}{dt} + \frac{\partial V}{\partial a} \frac{da}{dt} \right] \bigg/ \oint dt$$

$$= \oint dt \frac{\partial V}{\partial a} \frac{da}{dt} \bigg/ \oint dt \quad \text{using (4.1)} \tag{4.7}$$

Also,

$$\frac{\partial J}{\partial a} = \oint \frac{-dx}{\sqrt{[2(\bar{\varepsilon} - V)/m]}} \frac{\partial V}{\partial a} = -\oint \frac{dx}{v} \frac{\partial V}{\partial a} \tag{4.8}$$

Therefore, substituting (4.6), (4.7) and (4.8) into (4.5) gives

$$\frac{dJ}{dt} = \oint dt \frac{\partial V}{\partial a} \frac{da}{dt} - \oint \frac{dx}{v} \frac{\partial V}{\partial a} \frac{da}{dt}$$

$$= 0 \tag{4.9}$$

This example indicates that slow variations in the shape of the potential well do not change the value of J. However, the proof is not rigorous and several questions remain. For example, variations in the end points of the integral as $V(x, a)$ is changed have been ignored, and no estimate has been made of the error introduced by using an average $\bar{\varepsilon}$ for energy in defining the action integral. Nevertheless, the example suggests that J should remain nearly constant, even if large changes occur in V and $\bar{\varepsilon}$, provided that $a(t)$ is nearly constant during a complete period of the motion. A more rigorous analysis confirms that J is an adiabatic invariant and remains nearly constant for large alterations in $V(x)$ if these changes are made infinitely slowly.

The more formal Hamilton–Jacobi theory defines action-angle variables for periodic motion. If p_i and q_i are the momenta and its conjugate coordinate, then $J_i = \oint p_i \, dq_i$, where the integral is taken over the periodic orbit.

In the case of a charged particle in a magnetic field, the canonical momentum $\mathbf{P}$ is

$$\mathbf{P} = m\mathbf{v} + q\mathbf{A} = \mathbf{p} + q\mathbf{A} \tag{4.10}$$

where $\mathbf{A}$ is the vector potential of the magnetic field. The adiabatic invariants of the particle motion are then given by integrals of $\mathbf{P}$ over the appropriate periodic orbits. For charged particles in the geomagnetic field, three periodicities are readily apparent. These cycles correspond to the rapid gyration about the field lines, the north–south oscillation between magnetic mirroring points and the slow longitudinal drift about the Earth. Calculating the action integral associated with each of these periodicities leads to the three adiabatic invariants of the particle motion.

First adiabatic invariant

The so-called first adiabatic invariant is obtained by integrating $\mathbf{P}$ from equation (4.10) around the gyration orbit, where $d\mathbf{l}$ is an element of the particle path around the orbit.

$$\begin{aligned}
J_1 &= \oint [\mathbf{p} + q\mathbf{A}] \cdot d\mathbf{l} \\
&= p_\perp \cdot 2\pi\rho + q \oint \mathbf{A} \cdot d\mathbf{l} \\
&= p_\perp \cdot 2\pi \frac{p_\perp}{Bq} + q \oint \nabla \times \mathbf{A} \cdot d\mathbf{S}
\end{aligned} \tag{4.11}$$

where $d\mathbf{S}$ is an element of the area enclosed by the path. Therefore,

$$\begin{aligned}
J_1 &= \frac{2\pi p_\perp^2}{Bq} + q \oint \mathbf{B} \cdot d\mathbf{S} \\
&= \frac{2\pi p_\perp^2}{Bq} - qB\pi\rho^2 \\
&= \frac{2\pi p_\perp^2}{Bq} - \frac{\pi p_\perp^2}{Bq} = \frac{\pi p_\perp^2}{qB}
\end{aligned} \tag{4.12}$$

The second term in (4.12) is negative because $d\mathbf{S}$ as defined by the particle orbit points in the opposite direction to $\mathbf{B}$.

Rather than the above expression for J_1 the first invariant is taken to be $p_\perp^2/2m_0B$, which is equal to J_1 except for constant factors. The quantity

$$\mu = \frac{p_\perp^2}{2m_0B} \tag{4.13}$$

is often called the magnetic moment since in the non-relativistic limit it is equal to the current around the particle orbit times the area of the loop.

$$\mathcal{M} = I \cdot S = \left(\frac{v_\perp}{2\pi\rho}\right)q \cdot \pi\rho^2$$

$$= \tfrac{1}{2}q\frac{v_\perp m v_\perp}{Bq}$$

$$= \tfrac{1}{2}\frac{p_\perp^2}{mB}$$

The constancy of μ or of $p_\perp^2/B$ can be shown directly for simple geometries. For example, consider a particle in circular motion in a uniform field which increases with time. (See Figure 4.2.) The magnetic field is assumed to be symmetric about the center of the particle orbit so that the induced $\mathbf{E}$ is equal at all points in the orbit. If $\mathbf{B}$ is uniform and increases, the integral of Maxwell's equation over the area of the orbit gives

$$\oint \nabla \times \mathbf{E} \cdot d\mathbf{S} = \oint \mathbf{E} \cdot d\mathbf{l} = -\oint \frac{\partial \mathbf{B}}{\partial t} \cdot d\mathbf{S} = -\pi\rho^2\frac{\partial B}{\partial t} \tag{4.14}$$

The energy change in one revolution or in one gyroperiod τ_g is therefore

$$\Delta W = -q\oint \mathbf{E} \cdot d\mathbf{l} = q\pi\rho^2\frac{\partial B}{\partial t} \tag{4.15}$$

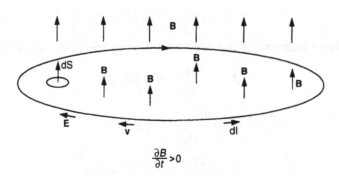

$$\frac{\partial B}{\partial t} > 0$$

Figure 4.2. Charged particle with velocity $\mathbf{v}$ gyrating in a uniform magnetic field which is increasing slowly with time. Magnetic moment of particle remains constant.

Hence

$$\frac{dW}{dt} = \frac{\Delta W}{\tau_g} = q\pi\rho^2 \frac{\partial B}{\partial t} \cdot \frac{Bq}{2\pi m}$$

$$= \frac{p_\perp^2}{2mB} \frac{\partial B}{\partial t} \tag{4.16}$$

Also,

$$\frac{dW}{dt} = \frac{d}{dt}(\gamma m_0 c^2) = m_0 c^2 \frac{d\gamma}{dt} \tag{4.17}$$

where

$$\frac{d\gamma}{dt} = \frac{d}{dt}\left[1 + \frac{p_\perp^2}{m_0^2 c^2}\right]^{1/2}$$

$$= \frac{1}{2m_0^2 c^2 \gamma} \frac{dp_\perp^2}{dt} \tag{4.18}$$

Equate (4.16) and (4.17) using (4.18) for $d\gamma/dt$ to obtain

$$\frac{1}{B}\frac{\partial B}{\partial t} = \frac{1}{p_\perp^2}\frac{dp_\perp^2}{dt}$$

Therefore

$$\frac{p_\perp^2}{B} = \text{constant} \tag{4.19}$$

and it follows that $\mu = p_\perp^2/2m_0 B$ is also constant.

The adiabatic or slow change constraint enters with the assumption that the orbit is circular. In fact, the gyration radius decreases as the acceleration takes place so the circular radius is only an approximation, valid for small changes in B during a single revolution.

The expression for the magnetic moment or first adiabatic invariant occurs naturally in many of the equations for particle motion. The mirror force equation (2.31) is given by

$$F_\parallel = -\tfrac{1}{2}qv_\perp\rho\nabla_\parallel B = -\frac{\mu}{\gamma}\frac{\partial B}{\partial s} \tag{4.20}$$

This equation is familiar as it describes the force on a dipole magnet of moment μ/γ in an inhomogeneous magnetic field. When the dipole field opposes the applied field, as is always the case when the magnetic moment is produced by a particle circling in the applied field, the force is repulsive, as given in equation (4.20).

Although the proof of the invariance of $p_\perp^2/B$ or $p_\perp^2/2m_0 B$ was shown for a time-dependent magnetic field, the invariance also applies if the particle moves into a region of different B, either by following a field line or by drifting across field lines. The motion parallel to field lines is particularly important, and it is in this case that the first invariant is most useful. Let α be the angle between the particle velocity and the magnetic field. This angle is customarily called the pitch angle:

$$\tan \alpha = \frac{v_\perp}{v_\parallel} \qquad (4.21)$$

Therefore, from equation (4.19),

$$\frac{p_\perp^2}{B} = \frac{p^2 \sin^2 \alpha}{B} = \text{constant} \qquad (4.22)$$

This equation is one of the most frequently used equations in radiation belt physics. Consider a particle whose pitch angle is α_{eq} at the equator where $B = B_{eq}$ (see Figure 4.3). As the particle moves along the magnetic field line towards the Earth, the field increases (see equation (3.21)). By (4.22), $p_\perp^2$ must also increase, and since p^2 is constant (in the absence of electric fields), $\sin^2 \alpha$ must increase. In the diagram $\alpha_2 > \alpha_1 > \alpha_{eq}$. When α reaches 90° the particle will be reflected, will return to the equator and will then repeat the trajectory in the opposite hemisphere. Equation (4.22) allows one to compute the pitch angle at any position of the trajectory, provided B is known at that position. It also specifies the magnetic field at the mirror point in terms of the field and the pitch angle at any other position. For example, if a particle has pitch angle α at B it will mirror at B_m where

$$\frac{p^2 \sin^2 90°}{B_m} = \frac{p^2 \sin^2 \alpha}{B} \qquad (4.23)$$

or $B_m = B/\sin^2 \alpha$ if $\mathbf{E} = 0$ and therefore $p = \text{constant}$.

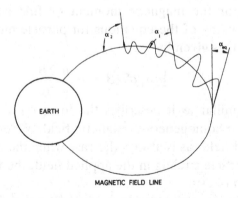

Figure 4.3. Conservation of magnetic moment results in an increase in pitch angle as the particle moves down the geomagnetic field line into the more intense field.

Note that the mirror point field is independent of particle momentum or charge and does not depend on the polarity of the magnetic field. The pitch angle at any point along a field line is given in terms of the equatorial values B_{eq}, α_{eq} or the mirror values B_m, α_m:

$$\sin \alpha(s) = \sqrt{\left[\frac{B(s)}{B_{eq}} \right]} \sin \alpha_{eq}$$

$$= \sqrt{\left[\frac{B(s)}{B_m} \right]} \qquad (4.24)$$

Equation (4.22) applies even when a parallel electric field $E_\parallel$ accelerates particles along a field line. Knowing a particle's energy, and therefore the momentum, as a function of B is sufficient to calculate B_m. The details of the helical trajectory through the electric field are not needed to find the mirror field.

At any location with field intensity B there will be few particles with $\alpha < \alpha_{LC} = \sin^{-1} \sqrt{(B/B_a)}$, where B_a is the field intensity at the top of the sensible atmosphere (~ 100 km). Particles with $\alpha < \alpha_{LC}$ will strike the atmosphere during each bounce and will be rapidly removed from the trapping region. The quantity α_{LC} is called the bounce loss cone angle. Because of the north–south asymmetry in the geomagnetic field the value of B_a on a given field line may be different at the two hemispheres. In such cases α_{LC} will be defined for the lower value of B_a and, therefore, for the larger value of α_{LC}.

Knowledge of the pitch angle as a function of position allows computation of such quantities as the time required to move between positions on a field line. For example, the time for a complete cycle of motion between mirror points s_m and s'_m is the bounce time τ_b where

$$\tau_b = 2 \int_{s_m}^{s'_m} \frac{ds}{v_\parallel(s)} = \frac{2}{v} \int_{s_m}^{s'_m} \frac{ds}{\cos \alpha(s)}$$

$$= \frac{2}{v} \int_{s_m}^{s'_m} \frac{ds}{\sqrt{\left[1 - \dfrac{B(s)}{B_m} \right]}} = \frac{2}{v} \int_{s_m}^{s'_m} \frac{ds}{\sqrt{\left[1 - \dfrac{B(s)}{B_{eq}} \sin^2 \alpha_{eq} \right]}} \qquad (4.25)$$

By changing the variable of integration from s to λ and using equations (3.19) and (3.21) the bounce time integral for a particle in a dipole field can be expressed in terms of the equatorial pitch angle α_{eq}, and the equatorial crossing distance R_0:

$$\tau_b = \frac{4R_0}{v} \int_0^{\lambda_m} \frac{\sqrt{(1 + 3\sin^2\lambda)}\cos\lambda\, d\lambda}{\left[1 - \frac{\sin^2\alpha_{eq}}{\cos^6\lambda}\sqrt{(1 + 3\sin^2\lambda)}\right]^{1/2}} \qquad (4.26)$$

The helical distance traveled during a complete bounce is

$$S_b = v\tau_b = 2\int_{s_m}^{s'_m} \frac{ds}{\sqrt{\left[1 - \frac{B(s)}{B_m}\right]}} \qquad (4.27)$$

Unfortunately, in a dipole field equation (4.26) cannot be integrated in closed form. However, an approximate formula good to about 0.5% is

$$\tau_b = 0.117\left(\frac{R_0}{R_E}\right)\frac{1}{\beta}[1 - 0.4635(\sin\alpha_{eq})^{3/4}]\text{ s} \qquad (4.28)$$

where R_0 is the distance from the center of the dipole to the equatorial crossing of the field line and $\beta = v/c$. Note the insensitivity of τ_b to the equatorial pitch angle. A particle mirroring near the equator has a bounce time of about half that of a particle which mirrors at the limiting distance near the dipole.

The constancy of μ is also useful in tracing paths for equatorially trapped particles ($\alpha_{eq} = 90°$). In the absence of electric fields, $\mu = $ constant requires the particle to drift along a contour of constant B, as expected since the gradient drift is perpendicular to ∇B. In the geomagnetic field which is more compressed on the sunward side the drift path will necessarily bring the particle closer to the Earth on the night side.

Graphs of the bounce periods for electrons and protons in the dipole approximation of the Earth's field are presented in Appendix B.

Second adiabatic invariant

The second mode of periodic motion of a geomagnetically trapped particle is the bounce motion between mirror points. If the longitudinal drift is small during a single bounce, the action variable associated with the bounce motion would be expected to be an invariant.

Returning to equation (4.10) for the canonical momentum, the action integral over a bounce is

$$J_2 = \oint(\mathbf{p} + q\mathbf{A}) \cdot d\mathbf{s} \qquad (4.29)$$

where $d\mathbf{s}$ is the element of length along a field line. The second term can be changed to a surface integral over the area **S** enclosed by the bounce path

$$\oint q\mathbf{A} \cdot d\mathbf{s} = q \int \nabla \times \mathbf{A} \cdot d\mathbf{S}$$

$$= q \int \mathbf{B} \cdot d\mathbf{S}$$

$$= 0 \tag{4.30}$$

since the integration path along the field line encloses a negligible area and no magnetic flux.

Therefore

$$J_2 = \oint \mathbf{p} \cdot d\mathbf{s} = \oint p \cos \alpha \, ds = \oint p_\| \, ds = \text{constant} \tag{4.31}$$

That J_2 in (4.31) is an adiabatic invariant can be seen from the following argument. From the equation of parallel motion for the non-relativistic case ($\gamma = 1$), equation (4.20) gives

$$F_\| = m\frac{dv_\|}{dt} = -\mu\frac{\partial B}{\partial s} \tag{4.32}$$

This equation can be integrated to give

$$\tfrac{1}{2}mv_\|^2 + \mu B = \mathcal{E}' \tag{4.33}$$

where $\mathcal{E}'$ is a constant of integration. Equation (4.33) is equivalent to (4.2) with $\mathcal{E}'$ corresponding to total parallel energy and μB corresponding to a potential energy. The parallel velocity is then

$$v_\| = \sqrt{[2(\mathcal{E}' - \mu B)/m]} \tag{4.34}$$

If B varies slowly with time, either by an explicit field change or by the particle drifting on to different field lines, one can define an average value $\bar{\mathcal{E}}'$ by integrating equation (4.33) over a bounce period as was done in equation (4.3). The value of J_2 expressed in terms of $\bar{\mathcal{E}}'$ is then

$$J_2 = \oint p_\| \, ds = \oint \sqrt{\{2m[\bar{\mathcal{E}}' - \mu B(s, a(t))]\}} \, ds \tag{4.35}$$

where $a(t)$ denotes a parameter which allows B to change slowly with time. That is,

$$\frac{\partial B}{\partial t} \ll \frac{B}{\tau_b} \tag{4.36}$$

where τ_b is the bounce period. With $\mu B(s, a(t))$ taking the role of the potential $V(x, a(t))$ the argument for $J_2 = \text{constant}$ is the same as developed in equations (4.4)–(4.9).

The second adiabatic invariant is often called the integral invariant. It is usually designated by J rather than by J_2 and this convention will be followed in the remainder of the book. To remove the particle momentum from the definition and express the invariant of a location entirely in terms of the magnetic field geometry, a related quantity, I, is often used as the integral invariant coordinate. For a point in space, I is defined in terms of J for a particle of momentum p mirroring at that point:

$$I = J/2p$$

$$= \tfrac{1}{2}\oint \cos\alpha\,\mathrm{d}s \tag{4.37}$$

or

$$I = \int_{s_m}^{s'_m}\sqrt{\left[1 - \frac{B(s)}{B_m}\right]}\,\mathrm{d}s \tag{4.38}$$

Again, s_m and s'_m are the locations of the mirroring points along a field line.

The primary use of the second or integral invariant is to define drift paths and the surfaces mapped out by the bouncing and drifting particle. In an axisymmetric magnetic field where $\mathbf{E} = 0$, this surface will also be axisymmetric since the gradient and curvature drifts are everywhere perpendicular to $\mathbf{B}$ and $\nabla_\perp B$. If the Earth possessed this idealized field, a drifting particle would circle the Earth and return to the initial field line. In a distorted field it is not clear from the drift equations that the guiding center drift path is closed. However, the invariance of J_2 or I ensures that the particle will return to the original field line.

Figure 4.4 illustrates a drift path in an asymmetric field. A particle initially on curve 1 on the right-hand side will drift to curve 2 on the left-hand side and return to 1, mirroring at B_m in both northern and southern hemispheres throughout the drift. At each longitude there is only one curve between mirror values of B_m having the required J value, because at a given longitude the value of J for given B_m increases monotonically with distance from the Earth.

Although the magnetic field is more compressed on the sunward side, the drift shell is closer to the Earth on the night side. For the same (r, θ) B is smaller on the night side, and the particle will move closer to the Earth to keep J constant. This result is similar to that for equatorial particles discussed on page 44.

In a distorted field there is no requirement that particles initially on the same field line but having different pitch angles will follow identical drift paths. Consequently, in their longitudinal drift, particles which started on the same field line may trace out different shells before returning to the

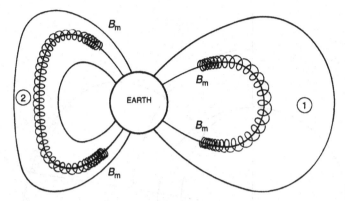

Figure 4.4. Trace of particle drifting and mirroring in the geomagnetic field. If the field is static, the particle will be reflected at fixed B_m and at each longitude will select the field line on which its motion between mirroring points conserves the second invariant.

initial line. This condition, called L-shell splitting, becomes important for field lines extending more than $\sim 4R_E$ from the Earth.

In the geomagnetic field distorted by anomalies and by the off-center dipole the mirroring altitude of a drifting particle will change with longitude, the altitude being lowest in the regions where the surface magnetic field is low. On each drift shell, the smallest pitch angle that still allows particles to drift completely round the Earth without striking the atmosphere defines a drift loss cone angle. At those longitudes where B_{max} is well above the atmosphere, particles mirroring below B_{max} can exist locally but are said to be in the drift loss cone. During their next drift orbit around the Earth, they will strike the atmosphere. The drift loss cone regions are in the 'magnetic shadow' of the Earth or its atmosphere and will contain only those trapped particles which have been diverted into that trajectory within their most recent drift period.

The rate at which trapped particles drift in longitude in a dipole field with $\nabla \times \mathbf{B} = 0$ is obtained by using equation (2.33) for the instantaneous drift velocity and averaging over a complete bounce period. From Figure 4.5 it is apparent that the instantaneous change in longitude ϕ is

$$\frac{\mathrm{d}\phi}{\mathrm{d}t} = \frac{V_\perp(r, \theta)}{r \sin \theta} \tag{4.39}$$

The change in ϕ during a complete bounce along a field line which crosses the equator at R_0 is

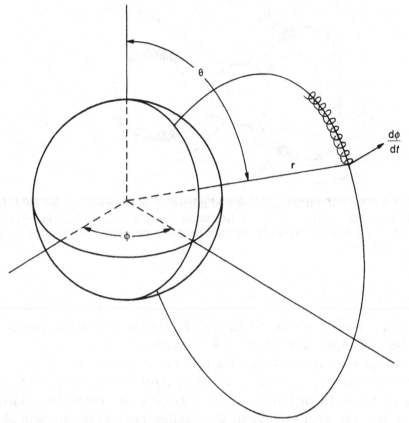

Figure 4.5. Coordinate system for calculating the longitudinal drift velocity of a trapped particle.

$$\Delta\phi = 4 \int_0^{s_m} \frac{V_\perp(\theta)}{R_0 \sin^3 \theta} \frac{ds}{v_\parallel} \tag{4.40}$$

where ds is the element of arc length along a magnetic field line measured from the equator and s_m is the distance along the field line from the equator to the mirror point. After changing the variable of integration from s to θ the time rate of change of longitude averaged over a bounce is

$$\left\langle \frac{d\phi}{dt} \right\rangle = \frac{\Delta\phi}{\tau_b} = \frac{4}{\tau_b} \int_{\theta_m}^{\pi/2} \frac{V_\perp(\theta)}{R_0 \sin^3 \theta} \left(\frac{ds}{d\theta} \right) \frac{d\theta}{v_\parallel} \tag{4.41}$$

Factors in the integrand of equation (4.41) can easily be computed for a dipole field. From equations (3.19) and (2.33):

$$\frac{ds}{d\theta} = R_0 \sin \theta [1 + 3\cos^2 \theta]^{1/2} \tag{4.42}$$

$$V_\perp = \frac{m}{qB^3}\left(\frac{v_\perp^2}{2} + v_\parallel^2\right)\mathbf{B} \times \nabla_\perp B \quad \text{(if } \nabla \times \mathbf{B} = 0) \qquad (4.43)$$

The parallel and perpendicular velocities can be expressed in terms of the equatorial pitch angle α_{eq} as

$$
\left.
\begin{aligned}
v_\perp^2(\theta) &= v^2 \sin^2 \alpha_{eq}\frac{(1 + 3\cos^2\theta)^{1/2}}{\sin^6\theta} \\[2mm]
v_\parallel^2(\theta) &= v^2\left[1 - \sin^2\alpha_{eq}\frac{(1 + 3\cos^2\theta)^{1/2}}{\sin^6\theta}\right]
\end{aligned}
\right\}
\qquad (4.44)
$$

Inserting these expressions into equation (4.41) results in an integral expression for the angular drift velocity:

$$
\left\langle\frac{\mathrm{d}\phi}{\mathrm{d}t}\right\rangle =
$$

$$
\frac{4}{\tau_b}\cdot\frac{3mvR_0^2}{qB_0R_E^3}\cdot\int_{\theta_m}^{\pi/2}\frac{\sin^3\theta(1 + \cos^2\theta)\left[1 - \frac{1}{2}\sin^2\alpha_{eq}\dfrac{(1 + 3\cos^2\theta)^{1/2}}{\sin^6\theta}\right]}{(1 + 3\cos^2\theta)^{3/2}\left[1 - \sin^2\alpha_{eq}\dfrac{(1 + 3\cos^2\theta)^{1/2}}{\sin^6\theta}\right]^{1/2}}\,\mathrm{d}\theta
$$

$$(4.45)$$

This equation cannot be integrated analytically, and values of the angular drift velocity and drift period $\tau_d = 2\pi/\langle\mathrm{d}\phi/\mathrm{d}t\rangle$ must be obtained numerically. For most purposes it is adequate to use an empirical fit which approximates the values of equation (4.45). An expression for the drift period accurate to $\sim 0.5\%$ is

$$\tau_d = \frac{2\pi qB_0R_E^3}{mv^2}\frac{1}{R_0}[1 - 0.3333(\sin\alpha_{eq})^{0.62}] \qquad (4.46)$$

This approximation can be simplified by collecting all constant factors to give

$$\tau_d = C_d\cdot\left(\frac{R_E}{R_0}\right)\frac{1}{\gamma\beta^2}[1 - 0.3333(\sin\alpha_{eq})^{0.62}] \qquad (4.47)$$

where

$$C_d = 1.557 \times 10^4 \text{ s for electrons}$$

and

$$C_d = 8.481 \text{ s for protons}$$

Note that as R_0 increases, the drift period decreases in spite of the larger drift path. Also as the particle velocity or energy increases, the drift period decreases. For non-relativistic particles such as protons below

50 MeV, the drift period is inversely proportional to energy. Equatorial particles drift more rapidly than those mirroring at higher latitudes, although this variation is not large.

Graphs of the drift periods of electrons and protons are plotted in Appendix B. Figures B.2 and B.3 give the drift periods for electrons and protons which mirror at the equatorial plane. The drift periods are plotted as a function of particle energy and equatorial crossing distance given by the parameter L, which is defined in the section on p. 53. For particles with equatorial pitch angles other than 90°, one must use Figure B.5. In Figure B.2 note the limiting value of τ_b as electron energy becomes relativistic. This condition occurs because the bounce period is simply the helical distance divided by the particle velocity. As the velocity approaches c, no further reduction in τ_b is possible.

Third adiabatic invariant

The third periodic motion of a geomagnetically trapped particle is the longitudinal drift about the Earth. In a static field, conservation of the first and second invariants will ensure that the particle returns to its original field line and will specify the field line occupied by the particle at each longitude. In a slowly changing magnetic field, μ and J are still conserved, but p may change in a way which depends on the details of the trajectory and the magnetic field changes. Hence, an additional constant of motion is needed to define trajectories in slowly changing magnetic fields.

The third invariant is derived as before by integrating the canonical momentum over the periodic trajectory:

$$J_3 = \oint (\mathbf{p} + q\mathbf{A}) \cdot d\mathbf{l} \qquad (4.48)$$

where in this case d**l** is the increment of longitudinal drift path, usually taken at the equator. The first term in (4.48) is neglected because the average momentum **p** in the direction of d**l** is small, $\mathbf{v}_d$ being orders of magnitude less than the actual particle velocity. Again, using Stoke's theorem,

$$J_3 = q \oint (\nabla \times \mathbf{A}) \cdot d\mathbf{S} \qquad (4.49)$$

where d**S** is an element of the surface enclosed by the equatorial drift path. Since $\nabla \times \mathbf{A} = \mathbf{B}$ the third invariant is

$$J_3 = q \oint \mathbf{B} \cdot d\mathbf{S} = q\Phi \qquad (4.50)$$

The quantity Φ is the magnetic flux enclosed by the drift path. Since the north–south oscillations are along field lines, the value of Φ does not depend on the latitude of dl as long as the drift path encompasses the Earth on the shell containing the guiding center trajectory. For this reason the third invariant is often called the flux invariant and is usually denoted as Φ, omitting the charge q.

Because of the magnetic singularity at the center of the Earth it is inconvenient to calculate the flux enclosed by the drift path. However, in the Earth's field the *net* flux inside the drift path is equal to the flux outside the path. Thus, the flux outside the integration path is usually computed to find Φ. In a dipole field the value of Φ for a particle at equatorial distance R_0 is

$$\Phi = \int_{R_0}^{\infty} B_0 \left(\frac{R_E}{r} \right)^3 2\pi r \, dr$$

$$= 2\pi B_0 \frac{R_E^3}{R_0} \tag{4.51}$$

Note that, as R_0 increases, the net flux enclosed decreases.

A non-relativistic proof of the invariance of Φ for a simple case is given below (see Figure 4.6). Assume that the particle is in the equatorial plane ($p_{\parallel} = 0$) and the magnetic field, which was initially B_1, slowly changes to a smaller value B_2. All vectors except $\mathbf{B}$ and $d\mathbf{S}$, the element of area in the surface enclosed by the drift path, lie in a plane. The particle will move

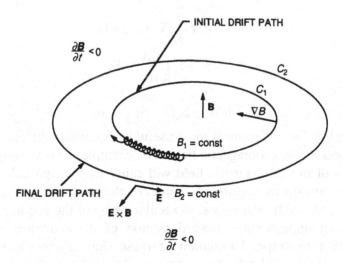

Figure 4.6. Conservation of the third adiabatic invariant. Magnetic flux enclosed by a particle drifting perpendicular to $\mathbf{B}$ and to ∇B remains constant during slow changes in the magnetic field.

from curve C_1 to C_2, being driven by the $\mathbf{E} \times \mathbf{B}/B^2$ force of the electric field induced by the changing magnetic field. Any increment to Φ resulting from $\partial B/\partial t$ is produced by a change in the flux density $\mathbf{B}$ within the original curve and by a change in the area enclosed by the drift path. Thus

$$\Delta \Phi = \Delta \Phi_B + \Delta \Phi_A \tag{4.52}$$

The change in Φ attributed to a change in $\mathbf{B}$ during a drift period τ_d is

$$\Delta \Phi_B = \tau_d \int_S \frac{\partial \mathbf{B}}{\partial t} \cdot d\mathbf{S} \tag{4.53}$$

where the integral is over the entire surface inside the particle drift path.

The increase in Φ associated with the change in the area enclosed by the drift path is

$$\Delta \Phi_A = \mathbf{B} \cdot \Delta \mathbf{S} = \tau_d \oint \mathbf{B} \cdot (\mathbf{V}_E \times d\mathbf{l}) \tag{4.54}$$

where $\mathbf{V}_E$ is the $\mathbf{E} \times \mathbf{B}/B^2$ drift velocity and $d\mathbf{l}$ is the element of drift path. Therefore,

$$\Delta \Phi_A = \tau_d \oint \mathbf{B} \cdot \left(\left(\frac{\mathbf{E} \times \mathbf{B}}{B^2} \right) \times d\mathbf{l} \right) \tag{4.55}$$

Expanding the triple vector product and noting that $\mathbf{B} \cdot d\mathbf{l} = 0$ results in

$$\Delta \Phi_A = \tau_d \oint \frac{\mathbf{B} \cdot (\mathbf{E} \cdot d\mathbf{l})\mathbf{B}}{B^2}$$

$$= \tau_d \oint \mathbf{E} \cdot d\mathbf{l}$$

$$= \tau_d \int_S (\nabla \times \mathbf{E}) \cdot d\mathbf{S}$$

$$= -\tau_d \int_S \frac{d\mathbf{B}}{dt} \cdot d\mathbf{S} \tag{4.56}$$

Thus

$$\Delta \Phi = \Delta \Phi_B = \Delta \Phi_A = 0 \tag{4.57}$$

The third or flux invariant is most useful in describing drift paths during slow changes in the geomagnetic field. For example, slow compressions or expansions of the geomagnetic field will cause trapped particles to move inward or outward as required to conserve the magnetic flux exterior to their orbits. Similarly, the very slow secular decay of the geomagnetic field results in an imperceptible inward motion of the radiation belts. The overall effect on trapped radiation of these slow changes is reversible; restoration of the field will return the particles to their original condition. Rapid changes in B, that is, $\partial B/\partial t \gtrsim B/\tau_d$, will cause permanent changes in Φ, as will be discussed in Chapter 6.

Geomagnetic coordinate system based on adiabatic invariants – the *L*-shell parameter

The lack of symmetry in the irregular geomagnetic field greatly compli-cates the tabulation of trapped radiation fluxes as a function of position, and in geographic coordinates a three-dimensional grid would be required for a complete description of flux values in space. Furthermore, a spatial coordinate system based on geographic coordinates loses the simplicity of the dipole formulas and does not lead to insights into the relationships of fluxes at different locations. What is needed is a coordinate system based on trapped particle motion which will have identical values for the coordinates of magnetically equivalent positions. By utilizing the near symmetry of the Earth's field to some extent the coordinate system would also eliminate the need for a longitude coordinate to describe the long-term trapped populations of particles.

The adiabatic invariants suggest such a system. The scalar value of the magnetic field is a useful coordinate since particles mirroring at a given B will mirror at the same value of B throughout their longitudinal drift. The second invariant, or, rather, its related function I, could be used as the second coordinate, since two positions in space with the same B and I values are magnetically equivalent from the standpoint of a trapped particle. Particles mirroring at a given value of B, I will drift around the Earth, mirroring at identical values of B and I in both hemispheres. Unfortunately, the quantity I is not an easy coordinate to interpret and does not vary linearly with any familiar variable. A more serious limitation is that it is not readily apparent from the values of B and I at several positions whether these locations lie near the same magnetic drift shell.

These difficulties are circumvented by a coordinate system devised by McIlwain. He recognized the convenience in a dipole field of the para-meter, R_0, the distance from the dipole center to the equatorial crossing or minimum B value of a field line. In a dipole, R_0 defines a field line as well as a drift shell and is readily visualized. During particle bounce and drift motion the particle remains on field lines having the same R_0. In a dipole field, if the B and I values of a location are known, the equatorial crossing point of the field line passing through that point can be deter-mined. Thus,

$$R_0 = f(I_D, B_D, \mathcal{M}_D) \tag{4.58}$$

where $f(I_D, B_D, \mathcal{M}_D)$ denotes a function of the dipole magnetic field value B_D, the integral invariant function I_D and the magnetic moment of the central dipole $\mathcal{M}_D$.

For the Earth's field, one then defines a new variable, L, in terms of the actual geomagnetic field. The L value of a location is based on values of B and I at that location calculated from the true, distorted geomagnetic field, but uses the same functional relationship relating the equatorial crossing distance to B and I for a dipole field. Thus,

$$LR_E = f(I, B, \mathcal{M}_E) \tag{4.59}$$

where $\mathcal{M}_E$ is the value of the dipole term for the Earth's field.

The variable L will be used as the second spatial coordinate. The distance LR_E is roughly equal to the distance from the center of the Earth's tilted, off-center, equivalent dipole to the equatorial crossing of the field line. Positions around the world having the same B and I will, by definition, have the same L values. Positions on the same field line in the distorted Earth's field will have very nearly the same values of L. Thus, a particle which bounces and drifts around the Earth will be very near to an $L = $ constant shell and can be assumed to follow an $L = $ constant path. In the distorted field L defines magnetic drift shells, and the value of L denotes the distance in Earth radii from the center of the equivalent dipole to the equatorial crossing for that drift shell.

It is frequently convenient to represent fluxes in the easily understood $r - \lambda$ polar coordinate system based on the B, L values of positions in the distorted Earth's field. For this purpose the coordinates r and λ of a point are defined implicitly by the B and L values of that point by

$$\left. \begin{array}{l} r = L \cos^2 \lambda \\[2mm] B = \dfrac{B_0}{r^3} \left(4 - \dfrac{3r}{L} \right)^{1/2} \end{array} \right\} \tag{4.60}$$

with r given in units of Earth radii. The values of B in terms of r and λ have the familiar dipole relationship of equation (3.15).

Computer programs which represent the Earth's distorted field usually include features which will compute the value of L for any position in space. Although the functional relationship of a position to its L coordinate is quite involved, in practice the conversion is routine.

A convenient way to interpret the B, L coordinate system for trapped particles is an follows. Imagine a dipole field with $\mathcal{M}$ equal to the dipole term in the Earth's field. If one removes trapped particles from the Earth's distorted field and places them in the imaginary dipole field at positions which preserve their mirroring B and J values, then one has a description of the radiation belts in the B, L coordinate system.

Problems

1. The one-dimensional, frictionless spring oscillator sketched below obeys the differential equation.

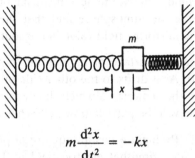

$$m\frac{d^2x}{dt^2} = -kx$$

   (a) If the maximum amplitude of oscillation is A, find the expression for the x coordinate as a function of time.
   (b) Find the expression for the adiabatic invariant of the system.
   (c) If, initially, the displacement is A and the springs are slowly strengthened to increase the frequency, what is the new amplitude when the frequency is double the initial value?

2. A satellite experiment is designed to produce artificial aurora by deflecting trapped electrons so that they enter the atmosphere. The satellite is in the equatorial plane at $R_0 = 3R_E$, and it generates electromagnetic fields which can deflect electrons:

   (a) Assuming that the Earth's field is a centered dipole and that electrons can be trapped only if they mirror above 100 km, what is the smallest pitch angle that a trapped electron can have at the satellite position?
   (b) To make sure that most of the electrons impact the atmosphere it is desired that the pitch angles at the 100 km altitude be 45°. For these electrons what must their pitch angle be at the equatorial plane? What is the minimum deflection angle that the satellite must supply to the trapped electrons to achieve this result?
   (c) If the satellite were crossing the $R_0 = 3R_E$ field line at a latitude of 45° what is the minimum required deflection?

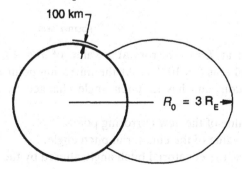

3. An electron of total momentum p is traveling along a field line between mirror points of strength B_m. In the northern hemisphere a weak D.C. electric field occurs, accelerating the electron downward to a new total momentum p_1. What is the magnetic field value at the new mirroring point? Assuming that the electron does not strike the atmosphere and that the electric field is maintained, what will be the mirroring field value in the southern hemisphere?

4. In a dipole field a charged particle of momentum p starts at $R_0 = 2R_E$ with second invariant $J = 0$. As it drifts to the other side of the Earth it encounters a weak electric field in the ϕ direction which slowly increases its momentum to $1.2p$. On what R_0 value will the particle now be? What will be its new J value?

5. Assume the particle in Problem 4 has an equatorial pitch angle of 45°. After being accelerated by the azimuthal electric field to $1.2p$, will the mirroring latitude be increased or decreased? Explain why.

6. A proton of momentum p is drifting around the Earth at $L = 2$ with an equatorial pitch angle of 90°. If the magnitude of the Earth's dipole moment slowly increases by 50%, at what distance from the dipole will the proton guiding center be at the end of this increase? What will its new momentum be?

7. An advanced civilization wishes to remove the radiation belts around its planet by constructing an electrostatic grid of wires in meridian planes extending several planet radii. As protons drift between these wires the $\mathbf{E} \times \mathbf{B}$ drift moves the protons outward. Repeated passages are intended to remove the protons to the region beyond the grids. Why won't this work using constant voltages on the grids? Would it work using time-dependent voltages? (Assume that the potential differences are much less than the energy of the protons.)

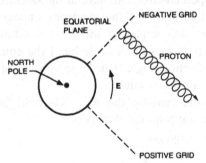

8. A trapped proton at $L = 4$ mirrors at a value of $B = 4 \times 10^{-5}$ T, while the equatorial field value is 5×10^{-7} T. At the mirroring point the proton collides with an oxygen atom and has its pitch angle changed by 5° with no loss in energy.
   (a) Find the B value of the new mirroring point.
   (b) Find the new value of the equatorial pitch angle.
   (c) How much was the equatorial pitch angle altered by the collision?

9. A non-relativistic proton with 90° pitch angle is drifting in crossed electric and magnetic fields starting at $x = y = 0$. The electric field $\mathbf{E}$ is in the x direction and is uniform. The magnetic field is directed into the paper and has a flux density which varies with y as $B = B_0 e^{\alpha y}$. The guiding center will move under combined $\mathbf{E} \times \mathbf{B}/B^2$ and gradient B drifts.

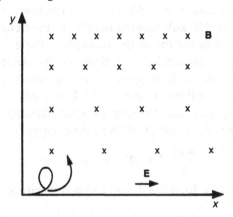

   Will the energy of the proton increase or decrease? If the initial momentum is p_0, what is the x, y position when the momentum has changed to p. (Note how readily the answer is obtained using conservation of the first adiabatic invariant.)

10. At $L = 4$, stably trapped electrons of 5 keV fill all pitch angles except for the loss cone (assume a dipole).
    (a) What is the value of the loss cone angle if there are no electric fields?
    (b) If an electric field directed upward from the Earth has an overall voltage drop of 1 keV between the equator at $L = 4$ and the atmosphere, what will be the equatorial loss cone angle for the 5 keV electrons?

11. A 5 MeV proton with second adiabatic invariant $J = 0$ is drifting on the $L = 2$ shell:
    (a) What must the energy be of a He^{++} ion (doubly charged He ion) so that it will drift at the same velocity on the same L shell? (Neglect relativistic effects.)
    (b) If the proton and helium ion are not confined to the equatorial plane but have the same pitch angle and the energies established in part (a), will they still drift at the same velocity?
    (c) Given the conditions in part (b), do the proton and He ion have the same J value?

12. A synchronous satellite in orbit at $L = 6.6$ is monitoring the loss cone while a TV camera on the ground is observing the atmosphere at the end of the field line passing through the satellite. (Assume a dipole field and neglect the thickness of the atmosphere.)

(a) At what magnetic latitude does the auroral observer locate his equipment in order to observe the light produced by particles intersecting the atmosphere?

(b) What is the angle of the loss cone as measured by the satellite at the equator?

(c) The satellite measures a sudden burst of electrons with energies ranging from 10 keV to 100 keV moving parallel to the field line. If the 100 keV electrons arrive at the top of the atmosphere 0.308 s after being detected by the satellite, how much later will the 10 keV electrons arrive?

13. A non-relativistic proton is trapped in the equatorial plane of a centered dipole field, and a uniform electric field **E** extends in the dawn to dusk direction. If the proton has velocity v_0 when crossing the noon–midnight meridian, show that the equation of its guiding center is

$$\frac{mv_0^2}{2}r^3 + q\mathbf{E}r^4\sin\phi = \text{constant}$$

where r is the distance from the dipole and ϕ is longitude.

5

Particle fluxes, distribution functions and radiation belt measurements

The specification of particle distributions

The preceding chapters have dealt with the motion of an individual particle in magnetic and electric fields. However, the measurement and the description of trapped radiation involves large numbers of particles distributed in space, energy and pitch angle. Hence, the concepts of a distribution function and flux have been introduced to describe the intensity and characteristics of a population of trapped particles.

The particle quantity used to describe trapped radiation intensities, energies and directionality is the particle flux. This quantity is the one most closely related to the output of most space radiation detectors. The differential, directional flux for a given location, direction and energy is the number of particles at energy E within unit dE which cross a unit area perpendicular to the specified direction within a unit solid angle in 1 s. Figure 5.1 illustrates this definition. If dA is the element of area, $d\Omega$ is the element of solid angle in the direction $\hat{\mathbf{e}}_\theta$, and dE is the energy interval at energy E under consideration, the number of particles with energies between E and $E + dE$ passing through dA in the direction $\hat{\mathbf{e}}_\theta$ within $d\Omega$ in 1 s is

$$dN(\mathbf{r}, E, \theta) = j(E, \theta)\, dA\, dE\, d\Omega \qquad (5.1)$$

where $j(E, \theta)$ is the differential, directional flux. (In this definition it is assumed that $d\Omega$ is sufficiently small that all particle velocities are nearly perpendicular to dA.) Note that the plane of area dA through which the particles must pass is perpendicular to $\hat{\mathbf{e}}_\theta$. The customary units of flux are $\text{cm}^{-2}\,\text{s}^{-1}\,\text{str}^{-1}\,\text{keV}^{-1}$, although the energy may be in MeV, keV or eV, depending on the energy range of the particles being described.

In the magnetosphere the most convenient reference direction is the geomagnetic field vector. Hence, the flux direction is usually given by the

Figure 5.1. Illustration of the definition of particle flux in the $\hat{\mathbf{e}}_\theta$ direction.

particle pitch angle. When trapped particle fluxes are symmetrical about
the magnetic field line, it is not necessary to indicate the azimuthal
dependence of the flux; only the pitch angle (α) dependence is needed.
(At locations where the flux has large gradients, such as near the atmosphere or the magnetopause, fluxes will not have this symmetry and the
azimuthal variation is needed for a full description.) The pitch-angle
distribution of trapped radiation provides clues as to the origin of the
particles and their loss mechanisms, as will be described in more detail in
Chapter 7. A typical pitch-angle distribution of protons in the stable part
of the belt is shown in Figure 5.2. The directional flux is usually a
maximum at 90° and falls to nearly zero at the loss cone angle. The flux
within the loss cone is very small because these particles will strike the
atmosphere before the next reflection and will be removed. The trapped
flux is symmetric about 90° since particles at pitch angles of α and
$180° - \alpha$ represent the same population, before and after magnetic mirroring. On a given field line the distribution will vary with latitude since the
loss cone angle increases as the observer moves from the equator towards
the atmosphere. In Figure 5.2 the upper curve represents the distribution
of the integral, directional flux which might be observed in the equatorial
plane. The other curves represent flux distributions as seen by a detector

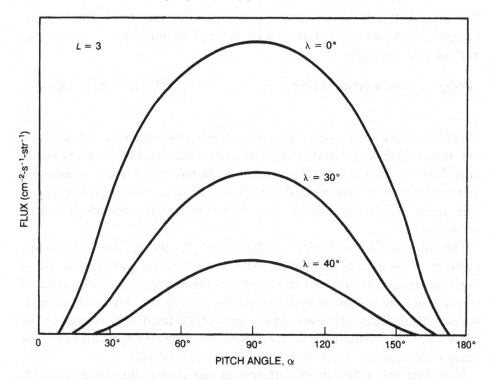

Figure 5.2. Typical pitch-angle distributions at three latitudes on the $L = 3$ field line. As latitude increases, the loss cone angle increases and the flux at 90° pitch angle decreases.

on the same $L = 3$ field line at latitudes of 30° and 40°. The fluxes at each pitch angle are reduced as λ increases, and the loss cone becomes larger with increasing latitude. Later in this chapter it will be shown how the trapped flux in the equatorial plane can be transformed to give flux values at arbitrary positions on the field line.

In this book the differential, directional flux will be designated by $j(\alpha, E)$. Other related flux definitions are

$$\text{Omnidirectional flux} = j(E) = \int_{\Omega} j(E, \alpha)\, d\Omega = \int_{0}^{\pi} j(E, \alpha) 2\pi \sin \alpha \, d\alpha$$
(5.2)

$$\text{Integral flux} = j(\alpha, E > E_0) = \int_{E_0}^{\infty} j(\alpha, E)\, dE$$
(5.3)

Although flux is usually plotted as a function of angle, the definition is in terms of solid angle. Hence, all angular integrals of fluxes are taken over solid angle as in the definition for omnidirectional flux in equation (5.2).

The integral flux is the flux above a given energy threshold. Similarly, an integral, omnidirectional flux can be defined by integrating over energy and solid angle to give

$$\text{Integral, omnidirectional flux} = j(E > E_0) = \int_{E_0}^{\infty} dE \int_{0}^{\pi} j(E, \alpha) 2\pi \sin \alpha \, d\alpha$$

$$(5.4)$$

Omnidirectional fluxes were often measured in the first years of radiation belt research since the early detectors did not distinguish the direction of arrival of the particles. Today, omnidirectional fluxes are a convenient unit with which to tabulate radiation levels in space since satellite engineers are primarily interested in the overall radiation level which will be encountered.

The integral flux, which is the flux above an energy threshold, is the quantity measured by detectors which respond to particles above a selected energy, determined either by the thickness of shielding material surrounding the sensor of by an electronic bias in the sensor electronics. Again, this flux definition was more appropriate for detectors used in the 1960s and for tabulating flux intensities for engineering purposes. The integral flux may be either directional or omnidirectional.

Note that the definition of differential, directional flux states that the 1 cm² area through which particles must pass is perpendicular to the particle velocity. This restriction is important in calculating such quantities as the flux of particles passing through a plane. For example, consider the flux entering the atmosphere in an auroral event in which the downward directed electron flux is isotropic, having filled the loss cone (see Figure 5.3). The downward moving integral flux is independent of α and is equal to j_0 cm^{-2} s^{-1} str^{-1}. We wish to calculate the number of electrons passing through an area of 1 cm² perpendicular to the magnetic field.

The number of particles at pitch angle α within a unit solid angle which pass through the unit area perpendicular to **B** is $j_0 \cos \alpha$. Thus the total number of downward moving particles passing through the unit area is

$$N = \int_{0}^{2\pi} d\phi \int_{0}^{\pi/2} j_0 \cos \alpha \sin \alpha \, d\alpha \qquad (5.5)$$

where ϕ is the azimuthal angle. Since j_0 is independent of ϕ and α, the number of particles crossing the unit area perpendicular to **B** is

$$N = j_0 \pi \qquad (5.6)$$

This value is half the value that one would obtain by the incorrect procedure of multiplying the directional flux j_0 by the downward directed solid angle 2π. The factor $\frac{1}{2}$ comes from the definition of flux, namely the

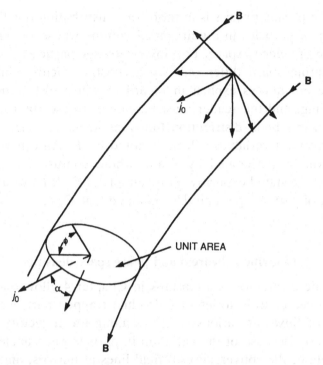

Figure 5.3. Calculation of the particle current across an area perpendicular to **B** in terms of the particle flux, j_o.

number of particles crossing a unit area *perpendicular* to the particle velocity. Since the unit area perpendicular to **B** in Figure 5.3 is not in general perpendicular to the particle direction, the cos α factor is needed in equation (5.5), resulting in the factor $\frac{1}{2}$.

Another common method of describing particle distributions is in terms of phase space density, a representation which has considerable theoretical importance. The phase space density is the number of particles per unit volume of six-dimensional space composed of the three orthogonal spatial dimensions and the three conjugate momenta. The importance of this coordinate system stems from the central role that phase space density plays in classical and statistical mechanics. For example, Hamilton's equations describe individual particle motion in phase space coordinates, and important theorems exist for these distributions. One of the most useful from the standpoint of space radiation is Liouville's theorem discussed in the next section.

If the particles are non-relativistic, the distribution is sometimes given in terms of velocity rather than energy or momentum. Much of the theory developed in plasma physics is in terms of a distribution function equal to the number of particles in a unit spatial volume whose velocities lie in a unit volume of velocity space. For non-relativistic particles, velocity space and momentum space differ only by a constant factor. Unfortunately, authors are not always explicit in regard to which distribution function they are using, and the reader must be careful to ascertain the variables involved. In this book distribution functions in phase space variables or adiabatic invariant variables will be denoted by F. All other distribution functions will be designated by f and where confusion is possible the variables will be stated explicitly. For example $f(E, J, L)$ would represent the number of particles per unit dE, dJ and dL at E, J, L.

Liouville's theorem and phase space densities

The adiabatic invariance relationships in Chapter 4 allow one to predict the guiding center trajectories of individual trapped particles. However, the values of fluxes at various positions along the trajectory are also of direct interest. Because of the variation in particle pitch angle during the bounce motion, the convergence of field lines at mirrors, and the irregularities in magnetic shells which are traversed during longitudinal drift, it might be expected that no simple relationship exists between the fluxes at different positions along the path of a particle.

Fortunately, a powerful theorem derived by Liouville can be applied to this question. This theorem will be proved for the case of one spatial dimension and then generalized for three spatial dimensions. Its use for trapped particle geometries will then be illustrated.

Liouville's theorem describes the density of particles in phase space, a space in which the coordinates are the usual spatial coordinates and the conjugate momenta. In the one-dimensional case illustrated here, let the spatial coordinate be q and the conjugate momentum be p (see Figure 5.4). The phase space density of particles is represented by $F(p, q, t)$, which is the number of particles which are at p, q per unit Δp and per unit Δq at time t. The number of particles in the interval q_1 to q_2 and p_1 to p_1 is given by

$$N(t) = \int_{p_1}^{p_2}\int_{q_1}^{q_2} F(p, q, t)\, dp\, dq \qquad (5.7)$$

In Figure 5.4 the number of particles entering the element $\Delta p \Delta q$ in unit time across the left side of the box is $F(p, q)\Delta p \dot{q}(q)$, evaluated at p, q, while the number of particles leaving the element on the right is

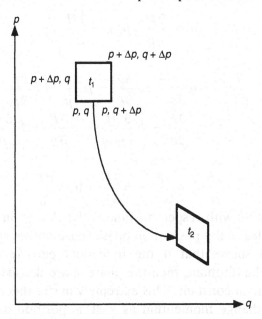

Figure 5.4. Path of a group of particles in one-dimensional phase space. The perimeter enclosing the particle group is distorted but maintains the same area.

$F(p, q + \Delta q)\Delta p \dot{q}(q + \Delta q)$. The difference in these two quantities is the rate at which particles accumulate in the phase space element due to currents across the vertical walls:

$$\frac{\partial F}{\partial t}\bigg|_q \cdot \Delta p \cdot \Delta q = F(p, q)\Delta p \dot{q}|_q - F(p, q)\Delta p \dot{q}|_{q+\Delta q}$$

$$\frac{\partial F}{\partial t}\bigg|_q = \frac{1}{\Delta q}\{F\dot{q}|_q - F\dot{q}|_{q+\Delta q}\}$$

$$= -\frac{\partial}{\partial q}(F\dot{q}) \quad \text{as } \Delta q \to 0 \tag{5.8}$$

Similar considerations apply for the top and bottom walls of the element $\Delta p \Delta q$:

$$\frac{\partial F}{\partial t}\bigg|_p = -\frac{\partial}{\partial p}(F\dot{p}) \tag{5.9}$$

The total rate of increase in the density of particles in the element is the sum of these two terms:

$$\frac{\partial F}{\partial t} = -\frac{\partial}{\partial q}(F\dot{q}) - \frac{\partial}{\partial p}(F\dot{p}) \tag{5.10}$$

Expanding the right-hand side and using Hamilton's equations:

$$\dot{q} = \frac{\partial H}{\partial p}, \quad \dot{p} = -\frac{\partial H}{\partial q} \tag{5.11}$$

gives

$$\frac{\partial F}{\partial t} = -\dot{q}\frac{\partial F}{\partial q} - F\frac{\partial \dot{q}}{\partial q} - \dot{p}\frac{\partial F}{\partial p} - F\frac{\partial \dot{p}}{\partial p}$$

$$= -\dot{q}\frac{\partial F}{\partial q} - F\frac{\partial^2 H}{\partial p\partial q} - \dot{p}\frac{\partial F}{\partial p} + F\frac{\partial^2 H}{\partial p\partial q}$$

or

$$\frac{\partial F}{\partial t} + \dot{q}\frac{\partial F}{\partial q} + \dot{p}\frac{\partial F}{\partial p} = 0 = \frac{\mathrm{d}F}{\mathrm{d}t} \tag{5.12}$$

The total derivative with respect to time is the change in the phase space density of particles as the position in phase space moves with the particles. Equation (5.12) shows that if the individual particle motions can be described by a Hamiltonian, then the phase space density along a dynamical path will remain constant. This extremely useful theorem applies even if the particles change momentum as well as position as a result of the forces.

In Figure 5.4 the particles which are in the phase space volume element t_1 at time t_1 will move to the box labeled t_2 at time t_2. Equation (5.12) indicates that the phase space density of the particles is unchanged. If the forces do not create or destroy particles, then the number of particles in the box at t_2 is equal to the number at t_1. This fact and the equality of particle densities requires that the volume of an element in phase space containing a group of particles is preserved as the particle distribution evolves with time. Therefore, the area of the box at t_2 equals that of the box at t_1, even though the shape at t_2 may be quite distorted.

Equation (5.10) can be generalized to three dimension, the result being

$$\frac{\mathrm{d}F(q_1, q_2, q_3; p_1, p_2, p_3; t)}{\mathrm{d}t} = \frac{\mathrm{d}F(\mathbf{q}, \mathbf{p}, t)}{\mathrm{d}t}$$

$$= \frac{\partial F}{\partial t} + \sum_{i=1}^{3}\frac{\partial F}{\partial q_i}\dot{q}_i + \sum_{i=1}^{3}\frac{\partial F}{\partial p_i}\dot{p}_i = 0 \tag{5.13}$$

In trapped radiation research the particle populations are usually described in terms of fluxes rather than phase space densities. Hence, to apply Liouville's theorem it is necessary to relate flux to phase space density. This relationship will be derived by describing the number of particles crossing an area per unit time in terms of flux and in terms of phase space density. Equating these two quantities will then give the desired relationship.

Consider an area dA whose plane is perpendicular to $\mathbf{v}$ and calculate the number of particles passing through dA s^{-1} in the direction of $\mathbf{v}$ within $d\Omega$ (see Figure 5.5).

In terms of flux $j(\theta, E)$ the number is

$$dN = dA \cdot j \cdot d\Omega \, dE$$

$$= dx \, dy \cdot j \, d\Omega \cdot \frac{p}{m} \, dp \tag{5.14}$$

In terms of phase space density, $F(\mathbf{p}, \mathbf{q})$, the number of particles crossing dA in one second is

$$dN = F \cdot dx \, dy \, dz \cdot dp_x \, dp_y \, dp_z$$

$$= F \cdot dx \, dy v \cdot p^2 \, d\Omega \, dp \tag{5.15}$$

where $dz = v \, (\Delta t = 1 \text{ s})$ is the length of the configuration space volume passing through dA in 1 s. The momentum space volume is the shell of thickness dp depicted in Figure 5.5.

Equating (5.14) and (5.15) gives

$$F(\mathbf{q}, \mathbf{p}) = j(\theta, E)/p^2 \tag{5.16}$$

Although the derivation given here is non-relativistic, the result is relativistically correct. The significance of this result combined with Liouville's theorem is readily apparent. Where p^2 is constant, the flux does not vary along a dynamical path. Therefore, in the steady state the observed differential, directional flux at any position along the dynamical path of a

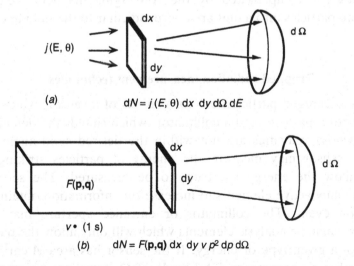

Figure 5.5. Description of a particle distribution in terms of flux (a) and phase space density (b).

bounce trajectory or a drift trajectory will be numerically equal. Thus, in the absence of time variations flux measurements at the same B, I values in the magnetosphere will give identical results. Even at different B and I values the directional fluxes will be identical if the two points are on a particle trajectory.

As an illustration of this flux equality, consider the pitch-angle distributions in Figure 5.2. These distributions are on the same field line but are at different latitudes and hence have different values of the magnetic field. At any latitude λ the directional flux at pitch angle $\alpha(\lambda)$ is equal to the equatorial flux at pitch angle α_{eq} where

$$\alpha_{eq} = \sin^{-1}(\sqrt{[B_{eq}/B(\lambda)]}\sin\alpha(\lambda)) \tag{5.17}$$

Thus, in Figure 5.2 the flux at $\alpha = 90°$, $\lambda = 40°$ is equal to the flux at $\alpha = 40.6°$, $\lambda = 30°$ and is equal to the flux at $\alpha = 21.6°$, $\lambda = 0°$. By applying Liouville's theorem and the conservation of the magnetic moment, the pitch-angle distribution at any point on the field line can be mapped from the distribution at the equatorial plane. However, the procedure cannot be reversed to derive a complete equatorial distribution from measurements made at higher latitudes since the high-latitude data do not include information on particles mirroring between the equator and the measurement point.

The explanation for the equality of flux at various positions on a field line is quite simple. As one moves from the equator into higher B, the solid angle containing a group of particles expands, diluting the flux. This effect is exactly compensated by the converging magnetic field which brings more particles into a unit area perpendicular to the field line.

Trapped radiation measurement techniques

In general a charged particle detector consists of a sensor, which records the impact of a particle, and a collimator, which shields the detector from particles whose velocities are not within the desired solid angle (Figure 5.6). The sensor may measure the energy of particles striking it and thereby allow the energy spectrum to be measured. The sensor may consist of many active elements to increase the information obtained from a detection event. The collimator or entrance aperture may contain electric or magnetic analysis elements which will only allow the passage of particles of a given type or energy. If the sensor has area A cm^2 and the collimator has an acceptance solid angle of Ω steradians pointing in the direction α with respect to **B**, the counting rate of particles with energies

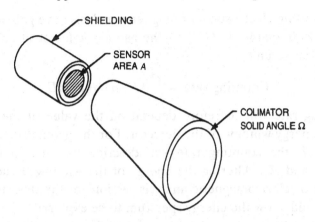

Figure 5.6. Generic charged particle detector consisting of a shielded sensor and an entrance collimator.

between E and $E + dE$ will be

$$\text{Counting rate} = j(\alpha, E)A\Omega\, dE \tag{5.18}$$

In equation (5.18) it is assumed that Ω is sufficiently small that all particles striking the sensor are moving perpendicular to A and that the variation of j over Ω is negligible. The more general case will be treated below. The combination $A\Omega = G$ is called the geometric factor of the detector. The geometric factor, whose units are expressed in cm^2 steradians, is a measure of the sensitivity of the detector. The larger the geometric factor the higher the counting rate in a given flux. The geometric factor should include a factor for the counting efficiency of the sensor. Here it will be assumed that all particles striking the sensor will initiate counts so that the efficiency is unity.

In designing experiments the geometric factor is selected to achieve a sufficiently large counting rate that statistical errors are small. If n is the number of counts recorded in an accumulation cycle and $n \gg 1$, the statistical error or standard deviation in the count value is $n^{\frac{1}{2}}$. The fractional error is therefore $n^{\frac{1}{2}}/n$, which becomes smaller as n increases. However, increasing n by enlarging either component of the geometric factor entails costs. If Ω is large, the angular resolution of the measurement suffers. A larger sensor area A increases the overall size and weight of the detector and collimator and increases the shielding weight required. Thus, the various parameters for the detector must be adjusted to optimize the experiment. Since fluxes vary greatly with position, angle and energy, it is not usually possible to design a detector that has good response under all the conditions that it will encounter in space.

Detectors using electrostatic or magnetic analysis have geometric factors which vary with energy. If $G(E)$ is the geometric factor at energy E, the counting rate is given by

$$\text{Counting rate} = \int_0^\infty j(E, \alpha) G(E) \, dE \qquad (5.19)$$

The counting rate will therefore depend on the value of the flux in the acceptance energy window of the detector. For the geometric factor shown in Figure 5.7, the counting rate will describe the flux in the interval between E_1 and E_2. Altering the energy of the window by adjusting the electrostatic or electromagnetic analysis sections of the detector will shift the window and allow the energy spectrum to be explored.

The energy spectrum can also be obtained directly using sensors whose output pulse amplitudes are related to the energy deposited in the sensor by the particle. The output pulses are fed to a pulse height analyzer which evaluates the energy of each particle, thus constructing an energy spectrum.

If the acceptance angle Ω is so large that the flux is not uniform over the aperture, integration is necessary to relate the counting rate to the flux. In Figure 5.8 let the vertical axis be in the direction of the magnetic field and the detector axis at angle θ with respect to the magnetic field. The detector of area A has a half-angle of acceptance of β and the acceptance solid angle is represented by a circle of radius β drawn on the unit sphere.

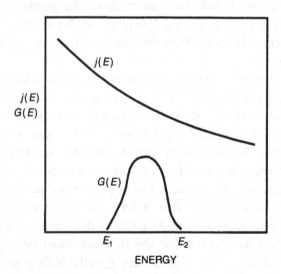

Figure 5.7. With an energy-dependent geomagnetic factor $G(E)$, the detector will sample the particle flux $j(E)$ at energies between E_1 and E_2.

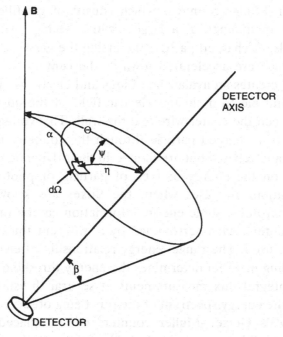

Figure 5.8. Geometry for calculating detector counting rates when the directional flux varies over the acceptance solid angle of the detector.

The flux is a function of pitch angle α, and integration of the flux over Ω is needed to obtain the counting rate. The differential of solid angle is

$$d\Omega = \sin \eta \, d\eta \, d\psi \tag{5.20}$$

The counting rate for an integral flux of $j(\alpha, E > E_{th})$ is

$$C.R. \, (E > E_{th}) = \int_0^{2\pi} \int_0^{\beta} j(\alpha, E > E_{th}) \cdot A \sin \eta \, d\eta \, d\psi \tag{5.21}$$

In the integrand α is a function of η, θ and ψ given implicitly by the cosine law for spherical triangles:

$$\cos \alpha = \cos \theta \cos \eta + \sin \theta \sin \eta \cos \psi \tag{5.22}$$

In the general case this integral must be evaluated numerically. In practice, one often expresses $j(\alpha, E)$ as a function of several parameters and adjusts the parameters to fit experimental counting rates.

Particle detectors

After the discovery of the radiation belts the techniques for measuring particle fluxes evolved rapidly, more or less in step with satellite technology and solid state electronics. The first measurements were taken using Geiger–Müller counters, which consist of gas-filled cylinders with center wires maintained at a large positive voltage with respect to the cylinder walls. A charged particle traversing the gas volume produces free electrons which are accelerated towards the center wire where the large electric field creates an avalanche of ions and electrons. The movement of these ions and electrons in the electric field of the counter reduces the voltage between the center wire and the wall. This voltage pulse indicates the passage of a charged particle through the counter. These devices are inexpensive and reliable but are limited in their dynamic range and give no information on the energy or type of particle or photon producing the initial ionization. In cases where the fluxes are known to be mostly electrons or protons, some energy information can be obtained if several identical counters, each surrounded by a different thickness of shielding material, are used. The range–energy relationship of electrons or protons in the shielding material determines the energy threshold for the detector, and these integral flux measurements at several thresholds give a crude estimate of the energy spectrum of the flux being observed.

In the 1960s Geiger–Müller counters were replaced by scintillation detectors and solid state sensors. In both of these types of particle detector the amplitude of the output signal, the light pulse from the scintillator or the current pulse from the solid state detector, is proportional to the energy deposited in the sensor material. For sensors large enough to stop the incoming particle, the pulse amplitude distribution of output pulses gives a direct measurement of the energy spectrum of the incident particles. Both scintillation counters and solid state sensors can sustain higher counting rates than can Geiger–Müller counters and therefore have greater dynamic ranges.

More elaborate detector systems having multiple sensors and discrimination elements are now common. A popular configuration is a two-element telescope consisting of a thin dE/dx sensor followed by a larger total energy detector. The thin element gives the energy loss per unit distance traversed in the material and the final detector measures the total energy deposited by the particle. These two quantities suffice to determine energy and particle identity, since dE/dx for a singly charged particle depends primarily on the particle's velocity. Knowing the velocity and energy leads to determination of the mass. Such telescopes easily distinguish electrons from ions, and in some versions can separate protons from helium or heavier ions. Added elements can be used to improve particle identification and discriminate against background. For example, a shield of active material, either a scintillator or solid state detector surrounding the counting system except for the entrance aperture, can be operated in anti-coincidence with the central detector and will reject signals from particles which do not enter through the aperture. Figure 5.9 illustrates an example of this type of particle spectrometer. The total energy sensor consists of a stack of silicon solid state detectors whose output signals are added to give a total energy pulse.

To detect low-energy particles which cannot penetrate the covers, light shields or dead layers of scintillators or solid state detectors, channel multiplier sensors are appropriate. The sensing element consists of a hollow tube whose interior is lined with a material having a low work function. A voltage of approximately 1 kV is applied between the ends of the tubes. An electron or ion entering the open end (negative potential) will strike the walls and initiate a cascade of electrons towards the positive end of the tube. At the positive end this electron cloud is collected, the charge is fed to an amplifier and the pulse is passed on to the detection logic. Since any charged particle or a photon can initiate a cascade the channel multiplier cannot discriminate between particle type or energy, and any particle analysis must be done prior to detection. Thus, ahead of the sensor are magnetic or electrostatic spectrometers or both. Combined magnetic and electrostatic analysis is a powerful technique since the former element separates particles by momentum and the latter by energy, thus leading to unambiguous determination of particle mass if the charge is known. In practice, these detectors vary the applied electric or magnetic fields in order to scan the energy and mass of the incoming particle flux.

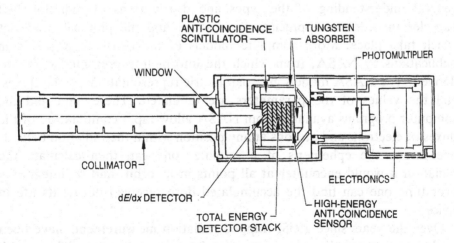

Figure 5.9. Highly directional particle spectrometer which will identify the particle species and measure the particle energy.

Arrays of channel multipliers, called channel plates, offer the possibility of recording an image of the incoming particle flux. Depending on the analysis elements which proceed the detector, the position of a particle when it strikes the channel plate can represent two characteristics, such as energy and angle of arrival or mass and energy. A relatively simple use of the channel plate is to place it behind a small aperture, with or without magnetic or electrostatic focusing, so that the position of detection on the plate records the direction of arrival of the particle. Such a detector gives high angular resolution for studying flux distributions with large angular gradients.

Solid state detectors are also used in arrays for detector focal planes. In addition to particle trajectory information which is given by the detector location, the energy of the particle is obtained if the detector is thick, and dE/dx is recorded if the detector is thin. Again, a combination of dE/dx, total energy and anti-coincidence shielding elements can be combined to yield information on the flux, particle type, energy spectra and angular distribution.

Trapped particle populations

Introduction

The purpose of this section is to outline the general characteristics of the Earth's radiation belts by describing the species, energy spectra, and spatial distributions of the trapped particles. It is important to have a general understanding of the types and distributions of radiation belt particles in order to appreciate their scope and the physical processes which take place. More complete models of the belts are available in publications by NASA, from which the information presented here was derived. The NASA models give parametric representations of the fluxes, and the values of the parameters are tabulated. These data and the computer programs available from NASA allow rapid computation of the fluxes of electrons and protons at any position within the radiation belts. If one knows the ephemeris of a satellite, one can then calculate the radiation it would encounter at all points in its orbit, and by integrating over time one can find the accumulated dose received during its life in space.

Over the years since 1958, trapped radiation measurements have been made by many satellites using a variety of detectors. In spite of this wealth of data, our knowledge of radiation belt characteristics is still incomplete.

Before 1975 many of the measurements were made with detectors with poor energy sensitivity and which were unable to discriminate between protons and electrons. Also, the geometric factors of some detectors were often uncertain, making it difficult to derive accurate flux values from the counting rates. Finally, the time variations of flux values in the belts are large, and the intensity and energy spectra vary with time of day, magnetic activity and solar cycle phase. Under these conditions it is difficult to evaluate average values for flux levels.

The particle populations described below have been derived by NASA from careful examination of data taken from a large number of independent experiments. Where necessary, interpolations and extrapolations were made to fill data gaps. While it is difficult to estimate errors, it is likely that errors of a factor of 3 are common, and errors of an order of magnitude may occur where large flux gradients exist. Also, because of the large time variations, the actual flux measured at any particular time may differ by an order of magnitude from the average values provided by these models.

In recognition of the need for better radiation belt models, the CRRES (Combined Release and Radiation Effects Satellite) was launched in 1990. This mission was equipped with modern particle detectors and obtained data for nine months as it traversed the belts in an eccentric, equatorial orbit. These data are now being incorporated into improved radiation belt models and should be available in a few years.

It is customary to describe the radiation belts as consisting of two zones, an inner zone for the region inside $L \approx 2.5$ and an outer zone for $L \gtrsim 2.5$. This terminology was adopted when the early detectors experienced counting rate maxima at about $L = 1.5$ and $L = 3.5$ with a minimum or 'slot' in between. This topology results from an intense flux of high-energy protons at $L = 1.5$ and a broad maxima of energetic electrons near $L = 3.5$. The region of the minima, $L = 2.0-2.5$, is termed the slot region, to designate the relative absence of energetic particles.

Radiation belt protons

Energetic protons are found throughout the region of the magnetosphere where the geomagnetic field will sustain trapping. Contours of the omnidirectional, integral flux of protons above $100\,\text{keV}$ are shown in Figure 5.10, where the intensity contours are plotted in the $r-\lambda$ coordinate system. The maximum flux occurs on the equatorial plane at about $L = 3.1$. On this plot, which depicts averages, the maximum flux along

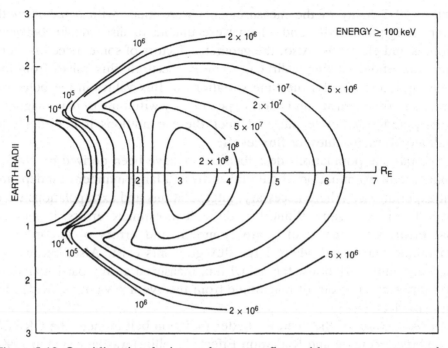

Figure 5.10. Omnidirectional, integral proton flux with energy greater than 100 keV. Based on data supplied by the National Space Science Data Center.

any *L* value will be found at the equator. The intensity contours become more closely spaced below *L* = 1.5, where the Earth's atmosphere becomes dense enough to remove the energetic protons.

Distributions are shown in Figure 5.11 for protons above 10 MeV and in Figure 5.12 for energies above 50 MeV. As the energy threshold increases, the flux maximum moves closer to the Earth. The increase in the average energy of trapped particles with decreasing *L* is a general characteristic of the radiation belts and will be discussed in Chapters 8 and 9. In these data, which are based on many experiments, the proton distribution displays a single maximum. Some experiments have reported transient secondary maxima, indicating that at times the proton distribution exhibits more structure than is shown here.

Proton energies extend up to several hundred MeV, making the proton fluxes the most penetrating of all trapped particles. Figure 5.13 illustrates the differential energy spectrum above 17 MeV. The extremely large energies, much larger than any electrical potentials occurring within the magnetosphere, suggest the existence of an energy source other than the electric and magnetic fields associated with the Earth.

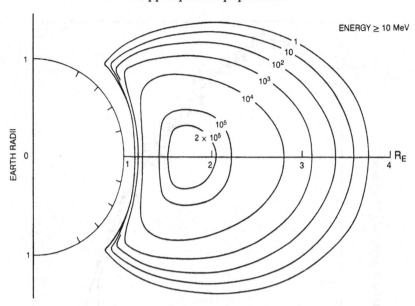

Figure 5.11. Omnidirectional, integral proton flux with energy greater than 10 MeV. Based on data supplied by the National Space Science Data Center.

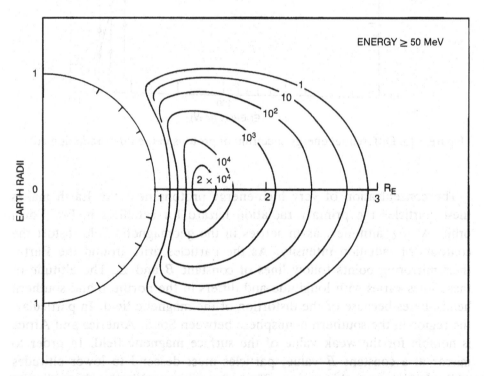

Figure 5.12. Omnidirectional, integral proton flux with energy greater than 50 MeV. Based on data supplied by the National Space Science Data Center.

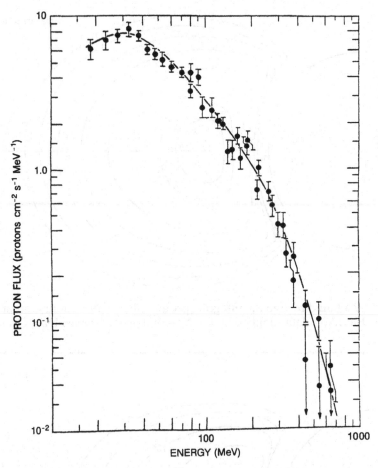

Figure 5.13. Differential energy spectrum of protons in the inner radiation zone.

The concentration of very high-energy protons near the Earth makes these particles the primary radiation hazard for satellites in low Earth orbit. At low altitudes, asymmetries in the geomagnetic field distort the contours of radiation intensity. As the particles drift around the Earth, their mirroring points follow lines of constant B and L. The altitude of these lines varies with longitude and differs in the northern and southern hemispheres because of the distortion of the magnetic field. In particular, the region in the southern hemisphere between South America and Africa is notable for the weak value of the surface magnetic field. In order to mirror at a constant B value, particles must descend to lower altitudes while drifting over this region. Thus, for a given altitude, the radiation intensity is much higher over the South Atlantic anomaly section than

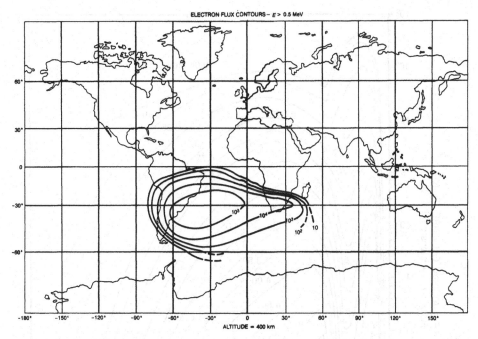

Figure 5.14. Radiation concentration at the South Atlantic anomaly. Isointensity contours of electrons above 0.5 MeV at an altitude of 400 km.

elsewhere. This effect is illustrated in Figure 5.14, which shows the radiation intensity of electrons with energies above 0.5 MeV at an altitude of 400 km. Contours of isointensity lines show the concentration of flux near the magnetic anomaly and indicate the importance of that region for satellite damage considerations.

Electrons

Energetic electrons occur throughout the Earth's trapping region and have energies extending up to several MeV. The fluxes exhibit time variations related to geomagnetic activity, and the average fluxes change during the solar cycle. As a result of this latter effect, different models are used for periods of solar maximum and solar minimum to bracket the expected average flux values. Because of the warping of the geomagnetic field by the solar wind, fluxes above $L = 5$ exhibit a local time variation that is incorporated in the models.

Figure 5.15 gives the equatorial values of integral, omnidirectional electron fluxes for a number of energy thresholds. The important features

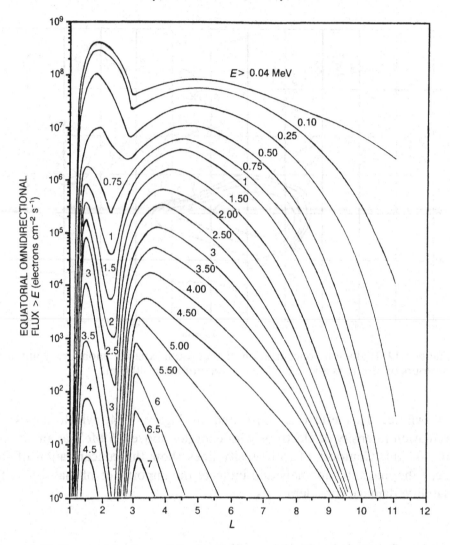

Figure 5.15. Equatorial values of the integral, omnidirectional electron flux above various energy thresholds. Based on data supplied by the National Space Science Data Center.

to note are the rapid decrease in flux with increasing energy, the concentration of high-energy electrons at the lower L values, and the slot region at $L \approx 2.5$. The slot, which is more pronounced for the higher-energy electrons, is believed to result from enhanced electron loss rates in this region.

Contour plots in $r-\lambda$ coordinates are given in Figures 5.16–5.18 for integral omnidirectional fluxes above 40 keV, 1 MeV and 5 MeV. In the

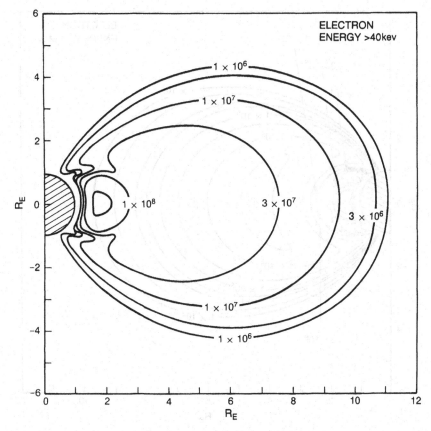

Figure 5.16. Integral, omnidirectional electron flux greater than 40 kev. Based on data supplied by the National Space Science Data Center.

first two diagrams the slot region is quite apparent. At L values of 3–5 the high-energy electrons are the most penetrating, and therefore the most damaging, component of the radiation belts.

Ions other than protons

In addition to electrons and protons the radiation belts contain significant components of heavier ions. The most abundant of these are helium and oxygen, although fluxes of carbon and nitrogen have also been identified. The most energetic of these heavy ions, with energies of a few MeV, have maxima in their radial distributions near $L = 3$ and have equatorial pitch-angle distributions which are sharply peaked near 90°. Therefore, the particles remain near the equatorial plane. At lower energies,

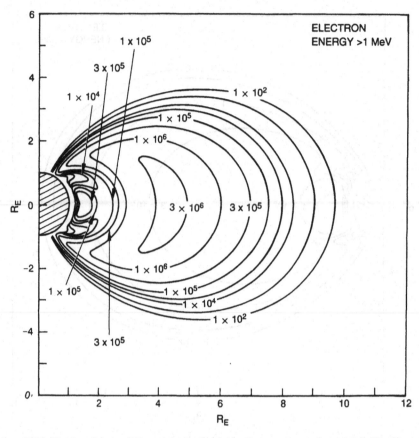

Figure 5.17. Integral, omnidirectional electron flux greater than 1 MeV. Based on data supplied by the National Space Science Data Center.

1–50 keV, oxygen and helium ions are quite abundant, particularly during magnetically active periods. There is strong evidence that these ions are drawn out of the atmosphere at high latitudes, and during substorms are accelerated and moved inward to $L = 3$ to 4. Occasionally, oxygen ions are the principal component of the ring current.

The presence of helium and oxygen ions trapped in the magnetosphere has led to important insights as to the origins of trapped particles and the physical processes responsible for the acceleration and trapping of ions. The singly charged oxygen ions must come from the Earth's ionosphere. Hence theories on the formation of the radiation belts must contain elements which permit the acceleration and trapping of the cold ionospheric plasma.

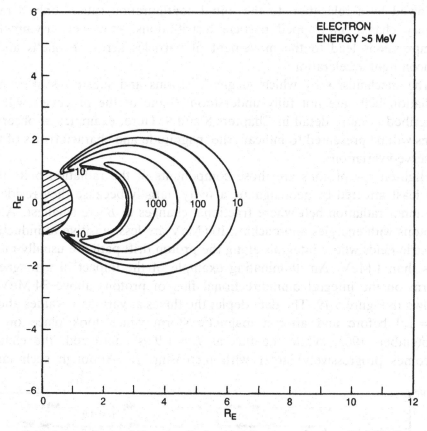

Figure 5.18. Integral, omnidirectional electron flux greater than 5 MeV. Based on data supplied by the National Space Science Data Center.

Time variations

All components of the radiation belts, both ions and electrons, exhibit time variations in flux intensity, energy spectra and spatial distributions. These variations are a major reason for the difficulty of mapping average values for satellite design purposes. Time variations are dominated by magnetic activity. Major magnetic storms produce changes throughout the trapping region, although the changes are largest in the outer zone. Magnetic substorms are more frequent but less severe disturbances in the geomagnetic field. During a substorm the configuration of the geomagnetic tail becomes less distorted and the induction electric fields modify the trapped particle distributions. Some of these changes are reversible. For example, symmetric compression of the geomagnetic field followed by a relaxation to the initial configuration causes only a temporary change in trapped particle distributions. However, asymmetric compressions lead to the movement of particles across L shells and a concomitant acceleration.

The mechanisms by which magnetic storms and substorms affect the radiation belts are not fully understood. Some of the processes will be described in more detail in Chapters 8 and 9. Here, examples of observations will be presented to indicate the magnitude and characteristics of the observed variations.

High-energy protons are those components of the radiation belts that are least affected by geomagnetic activity, largely because they reside in the inner radiation belt where fractional changes in B are smallest. Also, protons with energies approaching 100 MeV are less sensitive to induction electric fields whose integrals along the proton drift path are usually much less than 1 MeV. An illuminating example of the impact of a magnetic storm on the integral omnidirectional flux of protons above 34 MeV is shown in Figure 5.19. The data depict the fluxes at various L values above $L = 1.9$ before and after a magnetic storm which took place on 23 September, 1963. While the flux at $L = 1.9$ is unaffected, the change becomes progressively larger with increasing L. Although such large

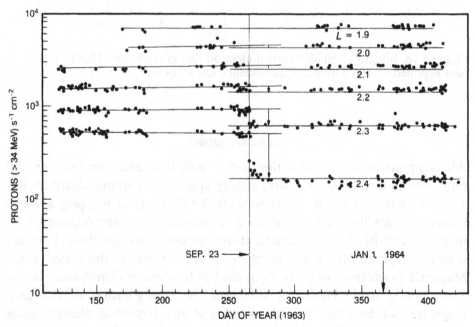

Figure 5.19. Sudden changes in high-energy proton fluxes caused by a major magnetic storm.

storms are rare, the long lifetimes of high-energy protons imply that they will experience a number of these storms.

At lower energies and higher L values the changes are more severe and complex. Figure 5.20 illustrates proton fluxes above several energy thresholds at $L = 4$. The storm which occurred on day 109 increased the flux at 2 MeV by an order of magnitude. In subsequent weeks following the storm, redistribution of the protons continued.

Electron fluxes are also susceptible to time variations, particularly in the outer radiation zone. The time behavior of 1 MeV electrons at L values between 3 and 5.5 is shown in Figure 5.21. The lowest panel of the diagram is a plot of the magnetic index, D_{st}, which indicates magnetic storms, and K_p, which is a measure of auroral activity at high latitudes. A decrease in D_{st} occurs during the main phase of a magnetic storm. Note

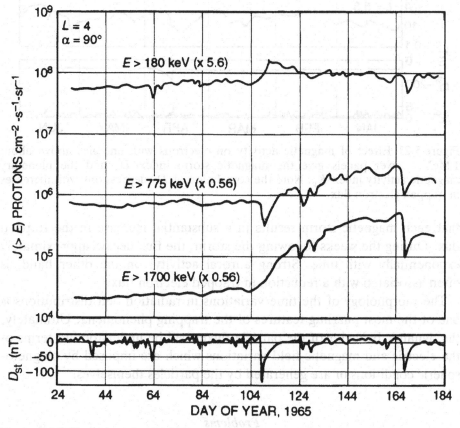

Figure 5.20. Variations in the integral, directional proton flux above three energy thresholds at $L = 4$. The lower panel gives the magnetic D_{st} index: a low value of D_{st} indicates a geomagnetic storm.

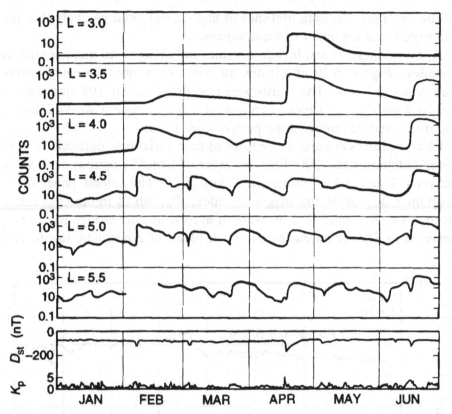

Figure 5.21. Effect of magnetic activity on electrons with energies above about 1 MeV. Lower panels give the magnetic storm index D_{st} and the planetary magnetic activity index K_p. Note the correlation of D_{st} depressions with increases in trapped electron flux.

that each magnetic storm results in a substantial increase in the trapped flux. During the weeks following the storm, the flux decays approximately exponentially with time. Strong auroral activity, on the other hand, is often associated with a reduction in trapped electron flux.

The morphology of the time variations in radiation belt distributions is one of the most puzzling features of the trapping phenomena. Ultimately, the dynamic behavior of the particle fluxes will be understood in terms of the electric and magnetic field variations which are imposed by magnetospheric conditions or are generated by the particles themselves.

Problems

1. An auroral scientist observes the intensity of the 391.4 nm line in a stable aurora and determines that 1 erg cm^{-2} s^{-1} of energy is being deposited in the

atmosphere. From the altitude of the luminosity he also estimates that the energy of the electrons entering the atmosphere is 5 keV. If he assumes that the flux entering the atmosphere is isotropic, find the integral, directional electron flux which would be measured in the loss cone by a satellite at the equator ($1 \text{ eV} = 1.6 \times 10^{-12}$ erg).

2. A directional proton spectrometer measures the proton flux at a point in space and obtains

$$j(\alpha) = 4 \times 10^2 \exp\left(-E/E_0\right) \sin^3 \alpha \quad (\text{cm}^{-2}\,\text{s}^{-1}\,\text{str}^{-1}\,\text{MeV}^{-1})$$

where α is the pitch angle and $E_0 = 10$ MeV:

(a) What is the omnidirectional flux above $E = 0$?
(b) What is the omnidirectional flux above 30 MeV?

3. A satellite designer finds that a CMOS integrated circuit installed in a satellite ready for launch will be damaged by energetic protons during the expected lifetime of the satellite. Because of weight restrictions she can only shield the circuit over 2π steradians. However, since the satellite is stabilized in attitude, she can place the 2π steradians of shielding in the most effective geometry. If the integral proton flux has an angular distribution of

$$j(\alpha) = C \sin^4 \alpha \quad (\text{cm}^{-2}\,\text{s}^{-1}\,\text{str}^{-1})$$

what is the fraction of the original flux that can be excluded? The satellite is to be in a circular orbit at the equator.

4. An engineer is designing a proton detector to measure protons of energy greater than 10 MeV using a scintillation counter with a pulse height threshold of 10 MeV. He wishes to have an angular resolution of ± 5 degrees and designs the collimator to this value. If the expected flux is 2×10^5 protons $\text{cm}^{-2}\,\text{s}^{-1}\,\text{str}^{-1}$ for $E > 10$ MeV and his count accumulation time is 200 ms, what is the area of the detector surface needed to achieve 1% statistical accuracy (standard deviation) in one accumulation interval?

5. Following a magnetic storm the integral electron flux above 100 keV measured on the equator over the Pacific at $L = 2$ is isotropic above a drift loss cone angle of $30°$, the cut-off resulting from the low field values at the South Atlantic magnetic anomaly.

 Find, for $L = 2$ at the longitude of the measurement given in the diagram below:

(a) The omnidirectional flux on the equator.
(b) The omnidirectional flux at a magnetic latitude of $30°$ (assume dipole field).
(c) When the distribution drifts to the Atlantic side of the Earth, will the equatorial omnidirectional flux be smaller, larger, or the same (assume no L-shell splitting)?

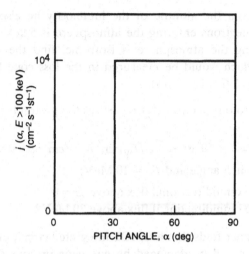

6. A magnetic mirror machine used for controlled fusion experiments has the configuration shown below. The field is symmetric about the z axis and is expressed in cylindrical coordinates. The z component is $B_z = B_0 e^{\alpha z^2}$. Charged particles with a radius of gyration much less than the dimensions of the machine will oscillate between mirroring points and slowly drift about the central axis:

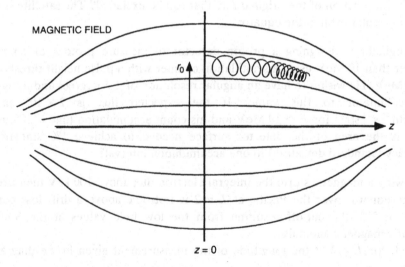

(a) Find the radial component of the field.
(b) A particle trapped between the magnetic mirrors at the ends of the configuration has its guiding center at a distance r_0 from the axis when passing $z = 0$. What is the third adiabatic invariant of the particle?

7. On the left-hand side of the diagram given below a differential, directional flux of protons

$$j(E, \alpha) = CE^{-2} \quad (\text{cm}^{-2}\,\text{s}^{-1}\,\text{str}^{-1}\,\text{keV}^{-1})$$

is uniform over pitch angles from 0 to 90°. The protons are accelerated by a potential difference V_0 and continue towards the right. Derive an expression for the flux on the right at pitch angle $\alpha = 0$ and energy E_0. Express the answer in terms of E_0, C, V_0 and q (charge on the proton).

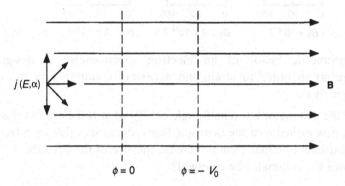

8. In the magnetic field configuration shown below a particle detector in B_1 measures the flux moving towards the right above some threshold energy and finds an angular distribution which is constant over the angular interval 0–90°. No measurements are made beyond 90°. Sketch the angular distribution of the flux from 0 to 180° at B_1, B_2 and B_3. Assume that all fluxes are time independent:

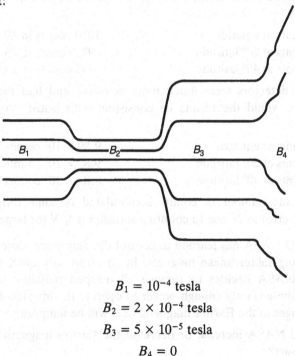

$$B_1 = 10^{-4} \text{ tesla}$$
$$B_2 = 2 \times 10^{-4} \text{ tesla}$$
$$B_3 = 5 \times 10^{-5} \text{ tesla}$$
$$B_4 = 0$$

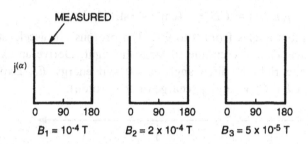

9. The geometric factor of an electron spectrometer is designed to be 0.05 cm² str in order to attain the necessary counting rate in an expected isotropic flux:

   (a) If the collimator has a half-angle of 10°, what is the area of the detector?
   (b) A new estimate of the isotropic flux reduces its value by a factor $\frac{1}{2}$. If the instrument builder cannot increase the size of the detector, to what angle must the collimator be changed?

10. A satellite with a unique guidance system is controlled to follow a magnetic field line at $L = 3$ from the equator to 40° in latitude. It carries an omnidirectional proton detector which accumulates counts for 10 s at the equator, at 20° latitude, and at 40° latitude:

   (a) Assuming that the loss cone angle is defined by the solid earth, is the detector response consistent with the flux being isotropic outside the loss cone?*

Counts at equator	1010 counts in 10 s
Counts at 20° latitude	972 counts in 10 s
Counts at 40° latitude	895 counts in 10 s

   (b) If the detectors were much more sensitive and had the following response, would the results be consistent with isotropy outside the loss cone?

Counts at equator	9.89×10^5 counts in 10 s
Counts at 20° latitude	9.8×10^5 counts in 10 s
Counts at 40° latitude	8.8×10^5 counts in 10 s

   * If N is the number of counts accumulated in some time interval, the probable error in N due to counting statistics is $\sqrt{N}$ for large N.

11. By 2020 AD NASA has learned to control the magnetic moment of the Earth by energizing subterranean magnets. In an effort to protect communication satellites, NASA decides to remove all trapped radiation below $L = 1.5$. However, there is only enough power to operate the magnets for a few hours so the changes to the Earth's magnetic field will be temporary:

   (a) Should NASA increase or decrease the Earth's magnetic moment for a few hours?

(b) By what factor should the Earth's magnetic moment be changed in that period? (Assume the Earth's field is a centered dipole with the field given by

$$\mathbf{B} = \frac{\mu_0}{4\pi} \frac{\mathcal{M}}{r^3}(-2\cos\theta\hat{\mathbf{e}}_r - \sin\theta\hat{\mathbf{e}}_\theta)$$

and neglect the atmosphere.)

12. At the equatorial plane a satellite observes an isotropic flux of electrons for $E > 5$ keV and $j(E < 5 \text{ keV}) = 0$:

$$j(\alpha, E) = 10^5 \, (E_0/E)^3 \text{ electrons (cm}^2 \text{ s str keV)}^{-1}$$

where $E_0 = 1$ keV:

(a) What is the value of the omnidirectional, integral flux above 5 keV?
(b) If the satellite is at $L = 2$, what is the loss cone angle (neglect the atmosphere)?
(c) What is the value of the power going into the atmosphere per cm^2 at the end of the field line?

6

Particle diffusion and transport

Introduction

The trapping properties of the geomagnetic field were described in preceding chapters. In particular, the guiding center equations and the adiabatic invariants were obtained, and it was found that energetic ions and electrons with appropriate initial position and velocity conditions were confined by the Earth's magnetic field. If the invariants were rigorously conserved, this confinement would be permanent; a trapped particle would remain trapped forever. However, rigorous conservation of the adiabatic invariants would also prevent other particles, such as those in the solar wind, in cosmic rays or in the ionosphere from ever becoming trapped. The radiation belts would then consist of only those particles which were injected in place by decaying neutrons or other radioactive particles.

There is a large body of experimental evidence showing that the adiabatic invariants are not conserved absolutely. Low-altitude satellites usually observe a flux of particles moving down the field lines and destined to be absorbed in the atmosphere. Also, the drift loss cone usually contains small but measurable fluxes of particles whose invariants must have been altered during their last drift period and which will be lost into the atmosphere during their current drift cycle. The observed time variations of both electron and ion fluxes also illustrate the frequent alteration of particle orbits. Although some of these variations can be attributed to reversible changes in the geomagnetic field and not to changes in the adiabatic invariants of the particles, most of the variations demand substantial modifications to the particle trajectories along with the injection of fresh particles and the loss of previously trapped ones.

The picture of the radiation belts is therefore one in which the magnetic container is an imperfect trap. Some defects are expected since the conditions required for the adiabatic invariants to hold rigorously are not always present. In the discussion of adiabatic invariants in Chapter 4 it was assumed that the magnetic and electric fields did not change appreciably during the cyclic motion of the particle. Field changes more rapid than any of the three cyclic motions associated with the three adiabatic invariants will lead to a change in the value of that invariant. In the magnetosphere a rich variety of electromagnetic and electrostatic waves are present with frequencies comparable to the gyration and bounce frequencies of the trapped particles. These waves will change the values of the corresponding adiabatic invariants and may remove the particles from the trapping regions. Magnetic activity, with its large-scale time variations in the electric and magnetic fields, also leads to a breakdown in one or more of the adiabatic invariants, usually the third invariant. The next three chapters will deal with changes in adiabatic invariants resulting from time variations in the fields and will show how field changes affect the particle distributions.

The equations for particle motion developed in the preceding chapters were deterministic in that the electric and magnetic fields were specified and the particle trajectories were then calculated. In treating the motion of charged particles in fluctuating fields a different approach is needed. In general, the time-dependent fields are not known precisely. Only the statistical properties, such as the power spectra and the geometrical pattern of the distortions, are known. It is therefore not possible to predict where an individual particle will be at some future time. One can only calculate the probabilities of the particle's behavior. When dealing with a large number of particles, these probabilities are quite adequate to describe the time evolution of the entire particle distribution.

Treatment of the time evolution of a distribution of particles whose trajectories are disturbed by innumerable small, random changes is by diffusion theory. However, this application of diffusion theory differs in concept from most other uses. In the usual diffusion applications, such as gases diffusing through porous media or neutrons diffusing through a moderator, the particle motion itself constitutes the diffusion. Any movement of the particles is tabulated as diffusion and is treated as such. If no diffusion occurs, the particles will remain in place. Diffusion of trapped particles is quite different. These particles may gyrate, bounce and drift around the Earth without actually diffusing. Only when one or more of their adiabatic invariants is altered can the particle be said to diffuse. Thus, the major motion is not associated with the diffusion processes to be studied, and one must remove this normal, adiabatic behavior from consideration in describing the diffusion process.

Diffusion equation

To illustrate the standard treatment of diffusion consider the one-dimensional case of particles diffusing through a porous column oriented along the x axis, as in Figure 6.1. In this case the distribution function $f(x, t)$ at time t is the number of particles in unit dx. Assume that at time $t = 0$ the distribution was $f(x, t = 0)$ and that there are no sources or sinks in the interval x_1 to x_2. The container has free boundaries at x_1 and x_2 where the particles can escape, so the distribution must vanish at those positions. The net current of particles diffusing across a position at x is proportional to the negative of the gradient of $f(x, t)$ at that point and is given by Fick's law:

$$\text{Current} = -D\frac{\partial f}{\partial x} \tag{6.1}$$

D being the diffusion coefficient. All the information on the physical mechanisms governing the diffusion process is contained in D. It is always positive and may of course be a function of x. The direction of net flow will be in the positive x direction if $\partial f/\partial x$ is negative. The net number of particles entering a section of length Δx in unit time by diffusing past x and $x + \Delta x$ is then

$$\Delta x\frac{\partial f}{\partial t} = D\frac{\partial f}{\partial x}\bigg|_{x+\Delta x} - D\frac{\partial f}{\partial x}\bigg|_x \tag{6.2}$$

In the limit of $\Delta x \to 0$

$$\frac{\partial f}{\partial t} = \frac{\partial}{\partial x}\left[D\frac{\partial f}{\partial x}\right] \tag{6.3}$$

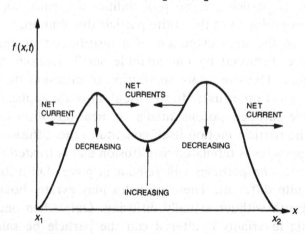

Figure 6.1. Impact of diffusion on a distribution of particles. Diffusion smoothes the distribution by decreasing the density at the peaks and increasing the density in the valleys.

Equation (6.3) is the diffusion equation which describes the time evolution of the distribution function. If D is independent of x, the instantaneous growth or decay of the particle density is proportional to the local value of $\partial^2 f/\partial x^2$. Where $f(x, t)$ is concave upward ($\partial^2 f/\partial x^2 > 0$), $\partial f/\partial t$ is positive and $f(x, t)$ will grow. Similarly, f will decrease at local maxima where $\partial^2 f/\partial x^2 < 0$. The diffusion process reduces peaks and increases valleys, thereby smoothing the distribution function (see Figure 6.1).

If $f(x, t)$ is known at some time, say $t = 0$, the value of f at a later time can be found by solving (6.3) as an initial value problem. In the general case where $D = D(x)$ the solution is obtained in the usual way by separation of variables. Let $f(x, t)$ be represented as the product of a function of x and a function of t:

$$f(x, t) = X(x)T(t) \tag{6.4}$$

Substituting (6.4) into (6.3) and dividing by $X(x)T(t)$ gives

$$\frac{1}{T}\frac{\partial T}{\partial t} = \frac{1}{X}\frac{\partial}{\partial x}\left[D\frac{\partial X}{\partial x}\right] \tag{6.5}$$

Because the left-hand side is independent of x and the right-hand side is independent of t, each term must be equal to a constant, which is designated $-\lambda_n$. Equation (6.4) can then be separated to give

$$\frac{1}{T}\frac{\partial T}{\partial t} = -\lambda_n \tag{6.6}$$

$$\frac{1}{X}\frac{\partial}{\partial x}\left[D\frac{\partial X}{\partial x}\right] = -\lambda_n \tag{6.7}$$

The time dependence is obtained by integrating (6.6) to give

$$T(t) = a_n\,e^{-\lambda_n t} \tag{6.8}$$

where a_n is a constant of integration. The x dependence of f is more difficult to obtain since equation (6.7) with $f(x_1) = f(x_2) = 0$ is an eigenfunction equation, having solutions for only specific values of λ_n. The function $X(x)$ will be expressed as a sum of eigenfunctions $g_n(x)$, where the g_n satisfy

$$\frac{\partial}{\partial x}\left[D\frac{\partial g_n}{\partial x}\right] = -\lambda_n g_n \tag{6.9}$$

with boundary conditions

$$g_n(x_1) = g_n(x_2) = 0$$

and are ordered such that $\lambda_1 < \lambda_2 < \lambda_3 \ldots$. These boundary conditions represent free escape or absorption at the boundaries where the particle density must vanish. If particles are reflected at the boundaries, different boundary conditions, such as $dg_n(x_1)/dx = dg_n(x_2)/dx = 0$ must be used. There are an infinite number of g_n, although in any practical problem n will be limited. Although the detailed shape of $g_n(x)$ depends upon $D(x)$, the general character follows the pattern illustrated in Figure 6.2 for the first three eigenfunctions. The lowest eigenfunction g_1 vanishes only at the boundaries and is always positive between x_1 and x_2. The next higher term becomes zero once in the interval x_1 to x_2, and subsequent g_n cross the horizontal axis $n - 1$ times between boundaries. Thus, higher n eigenfunctions have higher spatial frequencies and show more structure.

The eigenfunctions of (6.9) are orthogonal and can be normalized to give

$$\int_{x_1}^{x_2} g_n g_m \, dx = \delta_{nm} \tag{6.10}$$

where δ_{nm} is zero unless $n = m$, in which case $\delta_{nm} = 1$.

The general solution of (6.7) is a sum of the eigenfunctions, and the solution of (6.3) in the form of (6.4) is therefore

$$f(x, t) = \sum_n a_n e^{-\lambda_n t} g_n(x) \tag{6.11}$$

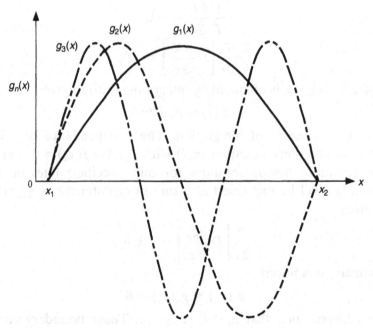

Figure 6.2. First three eigenfunctions of equation (6.9).

The constants a_n must be chosen to fit the initial conditions. If $f(x, t = 0)$ is the distribution at $t = 0$, each a_m can be obtained by multiplying both sides of equation (6.11) by g_m and integrating over x from x_1 to x_2. Because of the orthonormality of the eigenfunctions only the term containing g_m on the right-hand side survives. The coefficient a_m is then given by

$$a_m = \int_{x_1}^{x_2} f(x, t = 0)g_m(x)\,dx \qquad (6.12)$$

The magnitude of the coefficient a_m depends on the complexity of the x dependence of $f(x, t = 0)$. If the initial distribution has a single maximum, as does g_1, then a_1 will be large and coefficients for $n > 1$ will be smaller. Values of a_n for large n will be appreciable only if $f(x, t = 0)$ contains high spatial frequencies. Computing the a_n is equivalent to decomposing $f(x, t = 0)$ into its eigenfunction modes. The solution to the diffusion problem as represented by equation (6.11) is a sum of eigenfunctions with initial amplitudes, each decaying exponentially on a time scale of λ_n^{-1}. Since λ_n increases with n, the modes with high n decay more rapidly and the distribution eventually consists largely of g_1, which then decays as $\exp(-\lambda_1 t)$. The more rapid decay of higher modes leads to smoothing of the distribution function described earlier.

Diffusion in more than one dimension can be treated in a similar manner, although separation of the spatial variables is sometimes difficult. If one can choose a coordinate system in which one or more coordinates are constant on a boundary surface, the solution is obtained more easily. For example, spherical polar coordinates are appropriate for treating diffusion inside a ball. Cartesian coordinates are best for a rectangular block, and cylindrical coordinates simplify the treatment of diffusion in a right circular cylinder.

Particle diffusion in the radiation belts

As described previously, radiation belt particles undergo large-scale motion even if no diffusion is taking place. It is therefore necessary to cast the diffusion equation in a form which excludes normal adiabatic motion. The obvious coordinates for such a description are the adiabatic invariants themselves; a distribution function $f(\mu, J, \Phi)$ would remain unchanged if only adiabatic motion occurred. However, a distribution function in these coordinates is difficult to visualize and setting up a diffusion equation in these variables is not straightforward. One must resort to a general approach to diffusion theory which is valid for arbitrary coordinate systems. The Fokker–Planck equation provides such an approach as it is a prescription for deriving a diffusion equation in terms of the time rate of change in the coordinates of the particles.

A restricted form of the Fokker–Planck equation is derived as follows where $f(\mathbf{x}, t)$ is the number of particles per unit $\mathbf{x}$ at time t. The quantity $\mathbf{x}$ is a vector so that the following derivation is valid for diffusion in more than one dimension. Let $\Psi(\mathbf{x} - \Delta \mathbf{x}, \Delta \mathbf{x}, \Delta t)$ be the probability that a particle at $\mathbf{x} - \Delta \mathbf{x}$ will have its coordinates increased by $\Delta \mathbf{x}$ in time Δt. The characteristics of the mechanisms for producing the diffusion are contained in the probability Ψ. In the spirit of diffusion it is assumed that the processes driving the diffusion will change $\mathbf{x}$ by only a small increment during each elemental interaction. The probability $\Psi(\Delta \mathbf{x})$ will therefore be large only for small $\Delta \mathbf{x}$. In this concept the time interval Δt must be long compared to the time interval between the individual perturbations, yet short enough that major changes in the distribution function do not take place in Δt. At time $t + \Delta t$

$$f(\mathbf{x}, t + \Delta t) = \int d(\Delta x) f(\mathbf{x} - \Delta \mathbf{x}, t) \Psi(\mathbf{x} - \Delta \mathbf{x}, \Delta \mathbf{x}, \Delta t) \quad (6.13)$$

where the integral is taken over all values of the increment $\Delta \mathbf{x}$, although contributions to the integral will be small for large $\Delta \mathbf{x}$. Expand the left-hand side in a Taylor series about t and the right-hand side in a Taylor series about $\mathbf{x}$:

$$f(\mathbf{x}, t) + \frac{\partial f}{\partial t} \Delta t + \ldots = \int d(\Delta x) \left\{ f(\mathbf{x}) \Psi(\mathbf{x}, \Delta \mathbf{x}, \Delta t) - \left(\frac{\partial}{\partial x_i} (\Psi f) \right) \Delta x_i \right.$$

$$\left. + \frac{1}{2} \left(\frac{\partial^2}{\partial x_i \partial x_j} \Psi f \right) (\Delta x_i \Delta x_j) + \ldots \right\} \quad (6.14)$$

where the x_i are the components of $\mathbf{x}$. In this and the following chapters when an index is repeated within a single term, the term is to be summed over that index. The first term on the right-hand side of equation (6.14) is equal to $f(\mathbf{x}, t)$ since $f(\mathbf{x})$ is not a function of $\Delta \mathbf{x}$ and the integral over Ψ is unity. In the remaining terms on the right the order of differentiation and integration can be interchanged since $\Delta \mathbf{x}$ is not a function of x_i or x_j. With these changes equation (6.14) becomes

$$\frac{\partial f}{\partial t} \Delta t + \ldots = -\frac{\partial}{\partial x_i} \int d(\Delta x)(\Psi f)(\Delta x_i)$$

$$+ \frac{1}{2} \frac{\partial^2}{\partial x_i \partial x_j} \int d(\Delta x)(\Psi f)(\Delta x_i \Delta x_j) + \ldots \quad (6.15)$$

$f(\mathbf{x})$ is not a function of $\Delta \mathbf{x}$ and can be brought outside the integrals. If one denotes the averaged time rate of changes in the increments by brackets,

$$\langle \Delta x_i \rangle = \frac{1}{\Delta t} \int \Delta x_i \Psi(\mathbf{x}, \Delta \mathbf{x}, \Delta t) \, \mathrm{d}(\Delta \mathbf{x})$$

$$\langle \Delta x_i \Delta x_j \rangle = \frac{1}{\Delta t} \int \Delta x_i \Delta x_j \Psi(\mathbf{x}, \Delta \mathbf{x}, \Delta t) \, \mathrm{d}(\Delta \mathbf{x}) \qquad (6.16)$$

one obtains the Fokker–Planck equation as

$$\frac{\partial f}{\partial t} + \ldots = -\frac{\partial}{\partial x_i} \langle \Delta x_i \rangle f + \frac{1}{2} \frac{\partial^2}{\partial x_i \partial x_j} \langle \Delta x_i \Delta x_j \rangle f + \ldots \qquad (6.17)$$

For this equation to be useful, the higher-order terms in the Taylor series expansion must be negligible, a characteristic governed by the mechanisms driving the diffusion. If the incremental steps in the diffusion process are small, these higher-order terms are not important. The coefficients $\langle \Delta x_i \rangle$ and $\langle \Delta x_i \Delta x_j \rangle$ are sometimes called the first and second Fokker–Planck coefficients. In six-dimensional phase space there will be 42 of these coefficients. However, in the three-dimensional space of the adiabatic invariants the number is reduced to 12. Furthermore, only nine of these are independent since $\langle \Delta x_i \Delta x_j \rangle$ and $\langle \Delta x_j \Delta x_i \rangle$ are equal. In actual practice, some mechanisms affect only one variable; in that case the second coefficient is zero unless $i = j$. In the discussion leading to equation (6.20) it will be shown that the first Fokker–Planck coefficients can be expressed in terms of the second Fokker–Planck coefficients. Also, by selecting the coordinates cleverly, it is possible to further reduce the number of coefficients needed and simplify equation (6.17).

The transformation from six-dimensional phase space to the three-dimensional space of the adiabatic invariants represents a loss of information. The adiabatic invariant representation contains no information on the instantaneous phase of the particle, that is, its position in gyration, latitude or longitude. Such information is not needed if the particles are uniformly distributed in phase. In the remainder of this chapter it will be assumed that there is an efficient phase mixing mechanism which quickly restores phase uniformity after each elemental perturbation. The distribution function will therefore represent a value averaged over the phases of the variables.

The Fokker–Planck equation as derived here is linear in that the coefficients do not depend on $f(\mathbf{x}, t)$. This condition obtains if the probability of changes Ψ is independent of $f(\mathbf{x}, t)$. If one were considering mechanisms in which the particles act on themselves, for example by scattering from each other, this linearity would be lost and a more

complex approach would be needed. The great virtue of the Fokker–
Planck equation in this application is that it provides a prescription for a
diffusion equation, even if the coordinates and geometry are difficult to
interpret. One only needs to calculate the coefficients by applying equation (6.16) to obtain a valid equation. Although it is sometimes difficult to
obtain $\langle \Delta x_i \rangle$, this coefficient is not necessary, as will be seen next.

In Chapter 5 an equation for the evolution of the phase space distribution function $F(\mathbf{p}, \mathbf{q})$ was obtained in proving Liouville's theorem. This
equation, generalized to three spatial dimensions and three conjugate
momenta, is

$$\frac{\partial F}{\partial t} + \dot{q}_i \frac{\partial F}{\partial q_i} + \dot{p}_i \frac{\partial F}{\partial p_i} = \frac{\mathrm{d} F}{\mathrm{d} t} = 0 \tag{6.18}$$

From (6.18) it can be seen that if $\partial F / \partial q_i$ and $\partial F / \partial p_i$ are zero for all q_i and
p_i, then the time variation of F, namely $\partial F / \partial t$, must also be zero. Thus, if
the distribution is uniform in p_i and q_i, it will remain uniform and
constant.

If one uses phase space coordinates for the Fokker–Planck equation
(6.17) and postulates that the distribution $F_0(\mathbf{p}, \mathbf{q})$ is uniform in $\mathbf{p}$ and $\mathbf{q}$,
one obtains

$$\frac{\partial F_0}{\partial t} = 0 = -\frac{\partial}{\partial x_i} \langle \Delta x_i \rangle F_0 + \frac{1}{2} \frac{\partial^2}{\partial x_i \partial x_j} \langle \Delta x_i \Delta x_j \rangle F_0$$

$$= F_0 \frac{\partial}{\partial x_i} \left[-\langle \Delta x_i \rangle + \frac{1}{2} \frac{\partial}{\partial x_j} \langle \Delta x_i \Delta x_j \rangle \right] \tag{6.19}$$

The quantity in brackets must be a constant, independent of the mechanisms causing changes, and further considerations show that it is zero.
Thus, in phase space coordinates there is a relationship between the first
and second Fokker–Planck coefficients:

$$\langle \Delta x_i \rangle = \frac{1}{2} \frac{\partial}{\partial x_j} \langle \Delta x_i \Delta x_j \rangle \tag{6.20}$$

Although this relationship was derived for a uniform F_0, the Fokker–
Planck coefficients do not depend on the distribution function and the
result ((6.20)) is general. With equation (6.20) substituted for $\langle \Delta x_i \rangle$ in
equation (6.17) and neglecting the higher-order terms the Fokker–Planck
equation takes on the diffusion form

$$\frac{\partial F}{\partial t} = \frac{\partial}{\partial x_i} \left[\frac{\langle \Delta x_i \Delta x_j \rangle}{2} \frac{\partial F}{\partial x_j} \right] \tag{6.21}$$

where $\frac{1}{2} \langle \Delta x_i \Delta x_j \rangle = D_{ij}$ is the diffusion matrix.

The simplicity of equation (6.21) is a direct result of working in phase space variables. In any other set of variables, say y_i where the distribution $Y(\mathbf{y})$ is the number of particles per unit $\Delta \mathbf{y}$, the diffusion equation ((6.21)) would become

$$\frac{\partial Y(\mathbf{y}, t)}{\partial t} = \frac{\partial}{\partial y_i} \left[\frac{\langle \Delta y_i \Delta y_j \rangle}{2} \cdot \mathcal{J} \frac{\partial}{\partial y_j} \left(\frac{Y(\mathbf{y}, t)}{\mathcal{J}} \right) \right] \qquad (6.22)$$

where

$$\mathcal{J} = \frac{\partial(x_1, x_2, x_3)}{\partial(y_1, y_2, y_3)} \qquad (6.23)$$

is the Jacobian for transforming from phase space variables $\mathbf{x}$ to any other coordinates $\mathbf{y}$. There must exist a one-to-one mapping between the two coordinates systems, otherwise the Jacobian will be zero. Since $\mathcal{J}$ enters in both the numerator and denominator of (6.22), the constant factors in $\mathcal{J}$ cancel and need not be considered in making this transformation.

The most logical variables for radiation belt calculations are the adiabatic invariants μ, J and Φ. Since it is possible to transform from phase space variables to μ, J and Φ by a series of canonical transformations whose Jacobians are constants, equation (6.21) can immediately be written with the adiabatic invariants as independent variables. The transformed equation is

$$\frac{\partial F(\mu, J, \Phi)}{\partial t} = \frac{\partial}{\partial x_i} \left(D_{ij} \frac{\partial}{\partial x_j} F(\mu, J, \Phi) \right) \qquad (6.24)$$

where x_i with $i = 1, 2$ or 3 denotes the adiabatic invariants μ, J and Φ, and $F(\mu, J, \Phi) \, d\mu \, dJ \, d\Phi$ is the number of particles in the elemental volume $d\mu \, dJ \, d\Phi$. Transformations to other variables can be made using equation (6.22) and the appropriate Jacobian.

In treating diffusion in the radiation belts, a shrewd choice of coordinates based on the mechanism causing diffusion can greatly simplify the calculation. If an adiabatic invariant is not changed by the process considered, that invariant is a good one to use as a coordinate since the diffusion terms containing it will vanish. Possible choices of the three variables in addition to (μ, J and Φ) are (E, α_{eq}, L), (μ, B_m, L) and many others. In all cases it is essential that the Jacobian relating the adiabatic invariants to these new variables does not vanish.

For example, if one is dealing with diffusion in Φ only, the adiabatic invariants μ, J and Φ are a reasonable choice since μ and J are constant and the only diffusion coefficient required is $\langle (\Delta \Phi)^2 \rangle /2$. On the other hand, if only the pitch angles change in the diffusion process, then (E, α_{eq}, L) would be a good choice as the only term in (6.22) would be the one containing $\langle (\Delta \alpha_{eq})^2 \rangle$. In each of these cases the equation to be solved has only one term in the sum on the right-hand side.

Constructing the Jacobian for a given transformation is straightforward but sometimes tedious. The procedure is to express the adiabatic invariants in terms of the desired variables and then carry out partial differentiation and evaluate the Jacobian determinant. Each partial derivative operation is done keeping the other two independent variables constant. As an example, if one wishes to treat changes in Φ but prefers to use the set of variables (μ, J, L) (since L is more easily associated with experiments than Φ), the Jacobian is

$$\mathcal{J} = \begin{vmatrix} \dfrac{\partial \mu}{\partial \mu} & \dfrac{\partial \mu}{\partial J} & \dfrac{\partial \mu}{\partial L} \\[2mm] \dfrac{\partial J}{\partial \mu} & \dfrac{\partial J}{\partial J} & \dfrac{\partial J}{\partial L} \\[2mm] \dfrac{\partial \Phi}{\partial \mu} & \dfrac{\partial \Phi}{\partial J} & \dfrac{\partial \Phi}{\partial L} \end{vmatrix} = \begin{vmatrix} 1 & 0 & 0 \\ 0 & 1 & 0 \\ 0 & 0 & \dfrac{-1}{L^2} \end{vmatrix} = \frac{1}{L^2} \tag{6.25}$$

For diffusion in L the diffusion equation ((6.22)) then becomes

$$\frac{\partial f(\mu, J, L)}{\partial t} = \frac{\partial}{\partial L}\left[D_{LL} \frac{1}{L^2} \frac{\partial}{\partial L} (L^2 f(\mu, J, L)) \right] \tag{6.26}$$

where $D_{LL} = \langle (\Delta L)^2 \rangle / 2$.

A more complex example is the case of pitch-angle scattering in which α_{eq} is changed but the energy and L are kept constant. A good set of independent variables is $E =$ kinetic energy, L and $x = \cos \alpha_{eq}$. For these variables, the adiabatic invariants can be written as

$$\mu = \frac{E(E + 2m_0 c^2)(1 - x^2)}{2m_0 c^2 B_{eq}} \tag{6.27}$$

$$J = \frac{\sqrt{[E(E + 2m_0 c^2)]}}{c} 4 \int_0^{s_m} \cos \alpha(x, s)\, ds$$

$$= \frac{\sqrt{[E(E + 2m_0 c^2)]}}{c} L N_1(x) \tag{6.28}$$

$$\Phi = L^{-1} \tag{6.29}$$

In evaluating the Jacobian, note that B_{eq} in (6.27) is a function of L. Also, the derivative of $N_1(x)$ is obtained by expressing $\cos \alpha(x, s)$ in terms of x and differentiating the integral, leading to

$$\frac{dN_1(x)}{dx} = \frac{x}{(1 - x^2)} \{N_2(x) - N_1(x)\} \tag{6.30}$$

where

$$N_2(x) = \frac{4}{L} \int_0^{s_m} \frac{ds}{\cos \alpha} = \frac{\tau_b v}{L} \qquad (6.31)$$

is $1/L$ times the distance along the helix between mirror points. The resulting Jacobian, omitting constant terms, is

$$\mathcal{J} = \sqrt{[E(E + 2m_0 c^2)]} \, (E + m_0 c^2) L^2 x N_2(x) \qquad (6.32)$$

After canceling terms with no x dependence the diffusion equation for changes in $x = \cos \alpha_{eq}$ becomes

$$\frac{\partial f(E, x, L)}{\partial t} = \frac{\partial}{\partial x} \left[D_{xx} x N_2(x) \frac{\partial}{\partial x} \left(\frac{f(E, x, L)}{x N_2(x)} \right) \right] \qquad (6.33)$$

The remaining task is to relate the distribution function in whatever variables are chosen to the experimentally accessible quantities, namely the particle fluxes. This relationship can often be found directly using $f(x_1, x_2, x_3)$ to represent the number of particles in unit $dx_1 \, dx_2 \, dx_3$ and calculating the number of particles in an energy interval which cross a unit area in the equatorial plane at α_{eq} per unit time. This result can be used to obtain flux. A more systematic method is to use the relation (equation 5.16) between flux and phase space density. Since phase space density is equal (within constant factors) to the density of particles in the elemental volume defined by the coordinates (μ, J and Φ), the phase space density is related to the chosen distribution function by the Jacobian transforming that distribution to the space defined by the adiabatic invariants. Thus, the flux $j(E, a)$ is related to the phase space density by

$$j(E, \alpha) = p^2 F(\mathbf{p}, \mathbf{q}) \propto p^2 F(\mu, J, \Phi) \propto p^2 \frac{f(x_1, x_2, x_3)}{\mathcal{J}} \qquad (6.34)$$

For example, in the L diffusion case the particle flux is obtained directly from $f(\mu, J, L)$ and can be expressed in several ways:

$$j(E, \alpha) = p^2 f(\mu, J, L) L^2 = \frac{E(E + 2m_0 c^2)}{c^2} L^2 f(\mu, J, L)$$

$$= \frac{\mu 2 m_0 B_0}{L(1 - x^2)} f(\mu, J, L) \qquad (6.35)$$

The Jacobians for several selections of coordinates are given in Table 6.1.

The last selection of variables is suitable for treating simultaneous diffusion in pitch angle at constant energy and diffusion in L at constant mirror latitude. The scattering process only changes x and the diffusion in L only changes L. The variable z is fixed in both processes for equatorial particles and is nearly constant for $\alpha_{eq} \neq 90°$. The diffusion equation will therefore have only two terms and two diffusion coefficients, D_{LL} and D_{xx}.

Table 6.1. *Frequently used coordinates and corresponding Jacobians*

Variables	Jacobian is proportioned to
μ, J, L	L^{-2}
$L, E, x = \cos \alpha_{eq}$	$\sqrt{[E(E + 2m_0c^2)](E + m_0c^2)L^2 x N_2(x)}$
$\mu, L, \xi = (1 - x^2) = \sin^2 \alpha_{eq}$	$\mu^{1/2} \xi^{-3/2} L^{-5/2} N_2(\xi)$
$x, L, z = \dfrac{(E + 2m_0c^2)EL^3}{m_0^2 c^4}$	$L^{-5/2} x N_2(x) z^{1/2}$

As a generalization of Fick's law, the transformed equation (6.22) shows that the particle current across any coordinate y_i is given by

$$\text{current across } y_i = -\frac{\langle \Delta y_i \Delta y_j \rangle}{2} \mathcal{J} \frac{\partial}{\partial y_j}\left(\frac{Y(\mathbf{y}, t)}{\mathcal{J}}\right) \tag{6.36}$$

When only the diagonal elements $\langle (\Delta x_i)^2 \rangle$ of the diffusion matrix are not zero, the direction of flow will be given by the signs of the derivative factors. The value of this feature can be appreciated from the following one-dimensional example.

Using (μ, J, L) coordinates and treating diffusion in L, the derivative term in (6.36) for equatorial mirroring particles becomes (see equations (6.26) and (6.35))

$$\frac{\partial}{\partial L}(L^2 f) \propto \frac{\partial}{\partial L}\left(L^3 j\left(E, \alpha_{eq} = \frac{\pi}{2}, L\right)\right)_{\mu, J} \tag{6.37}$$

where the last relation in equation (6.35) is used to express flux in terms of $f(\mu, J, L)$. Because μ and x are constant for equatorial particles which diffuse in L, keeping μ and J constant, only the L dependence of $\mathcal{J}$ is retained.

A plot of $L^3 j(\alpha_e = \pi/2)$ as a function of L, where j is evaluated at constant μ (not constant energy), will immediately show whether particles are diffusing inwards towards the Earth or outwards towards the magneto-pause. This flow direction is determined entirely by the particle distribution and is independent of the mechanism causing the diffusion. The magnitude of the diffusion is, of course, proportional to the diffusion coefficient and is therefore influenced by the diffusion mechanism (see Figure 6.3).

It was stated earlier that a careful choice of coordinates could greatly reduce the difficulty of working with the general diffusion equation. Three

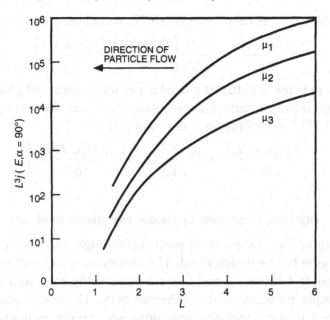

Figure 6.3. Illustration of how the particle flow direction is related to the derivative of the distribution function.

coordinates are necessary to describe a particle's condition (after averaging out the three cyclic phase variables) and the diffusion matrix therefore contains nine elements. This number can usually be reduced by choosing one, or at most two, variables which are not altered by the diffusion process. If two of the variables are constants, the diffusion matrix has only one non-zero element, and the diffusion equation is one-dimensional, similar to equation (6.3). Even if two coordinates are active during the diffusion, it may be possible to choose variables in which the changes are uncorrelated, in which case the off-diagonal terms $\langle \Delta x_i \Delta x_j \rangle i \neq j$ vanish.

Finally, the choice of coordinates can be influenced by the ease of calculation of the diffusion coefficients. The radial coordinate L is a convenient variable for evaluating diffusion across magnetic shells as the change in L by magnetic and electric field fluctuations is readily calculated (see Chapter 8). Diffusion in the pitch angles of trapped electrons by collisions with atmospheric atoms is simplified by using $x = \cos \alpha_{eq}$ as the basic variable, as will be shown in the next chapter.

Because of the special advantages of phase space coordinates many authors prefer to use a distribution function proportional to the density in phase space even if the coordinates in the diffusion equation are not phase space variables. In this case equation (6.22) would be

$$\frac{\partial F(\mathbf{y}, t)}{\partial t} = \frac{1}{\mathcal{J}} \frac{\partial}{\partial y_i} \left[\frac{\langle \Delta y_i \Delta y_j \rangle}{2} \mathcal{J} \frac{\partial F(\mathbf{y}, t)}{\partial y_j} \right] \tag{6.38}$$

where $F(\mathbf{y}, t)$ is the number of particles per unit volume of phase space at $\mathbf{y}$ and t. The diffusion equation for phase space density corresponding to equation (6.26) for L diffusion at constant μ, J is

$$\frac{\partial F(\mu, J, L)}{\partial t} = L^2 \frac{\partial}{\partial L} \left[D_{LL} \frac{1}{L^2} \frac{\partial F(\mu, J, L)}{\partial L} \right] \tag{6.39}$$

Injection of protons by cosmic ray albedo neutrons

In this chapter the diffusion of particles through the Earth's magnetic trapping region has been described. This diffusion, which will be treated in greater detail in Chapters 7 and 8, must occur if the solar and ionospheric plasmas supply particles to the radiation belts. However, one source of radiation belt protons and electrons does not require non-adiabatic motion. This source is the spontaneous decay within the trapping region of energetic neutrons produced by cosmic ray collisions with the Earth's atmosphere.

When a cosmic ray strikes the nucleus of an atmospheric atom, the products include high energy neutrons. Some of these neutrons escape the atmosphere immediately; others escape after further collisions. Thus, in the region around the Earth there is a flux of outward moving neutrons, a small fraction of which will decay within the magnetosphere. A neutron decays into a proton, an electron and a neutrino, the half life of the neutron being about 630 s. Because the proton mass is much larger than that of the electron or the neutrino, a proton from neutron decay will move initially with almost the same velocity as the parent neutron. If the decay takes place in the geomagnetic field and the newly born proton has a pitch angle outside the loss cone, the proton will be trapped. Electrons from the decay are emitted isotropically in the center of mass system and will also be trapped if their pitch angles are outside the loss cone. The method of calculating the proton source strength is sketched below for equatorially trapped protons.

Figure 6.4 illustrates, on an exaggerated scale, the gyration of a trapped proton in the equatorial plane at distance $L R_E$ from the center of the Earth. Each differential element of its path may be described by a vector $\mathbf{dr}$ whose sense is the same as the velocity of the proton. If $\mathbf{dr}$ extended backward intersects the Earth's atmosphere, then a neutron originating at

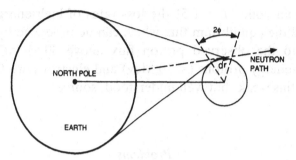

Figure 6.4. Geometry of the source of protons from decay of cosmic ray albedo neutrons.

that intersection point can decay while traversing the element dr and thus add a proton to the proton flux in dr. If the extension of dr does not intersect the atmosphere, injection is not possible in that element.

If $j_n(E, \alpha)$ is the differential, directional neutron flux in the direction dr, then the source of trapped proton flux at that position and direction due to neutron decays in a 1 cm^2 column of length dr is given by

$$\frac{dj_p(E, \alpha)}{dt} dr = j_n(E, \alpha)\frac{dr}{v} \cdot \frac{1}{\gamma\tau_n} \qquad (6.40)$$

where τ_n is the mean life of the neutron and γ is the relativistic time dilatation factor. The factor dr/v is the time a neutron of velocity v spends in an element of length dr.

The average flux increase over the gyration cycle of the proton is obtained by integrating both sides of equation (6.38) over a complete gyration, noting that $j_n(E, \alpha)$ vanishes when the backward extension of dr does not intersect the Earth. In the equatorial plane with gyration radius $\rho \ll R_E$ the geometry of Figure 6.4 shows that neutrons contribute to j_p only over the fraction ϕ_m/π of the gyration, where $\sin \phi_m = 1/L$. Thus, if j_n is independent of the zenith angle, the growth in proton flux is given by

$$\frac{dj_p(E, \alpha_{eq} = \pi/2)}{dt} = \frac{1}{\pi}j_n\left(E, \alpha_{eq} = \frac{\pi}{2} \right)\frac{1}{v\gamma t_n} \sin^{-1}\frac{1}{L} \qquad (6.41)$$

If $j_n(E, \alpha)$ is not independent of the zenith angle, this dependence must be included in the integration of equation (6.38). For the general case of injection off the equator, where neutrons emitted at all latitudes may contribute, the latitude dependence of the albedo neutron flux must also be included.

Extensive numerical calculations of the neutron decay injection show that the source of high-energy protons is extremely small. However, within the inner zone ($L \approx 1.5$) the loss rates of high-energy protons are very low, and the equilibrium flux which can be produced by this source is comparable to the observed proton flux above 50 Mev. The observed fluxes of low-energy protons (< 10 Mev) and electrons are too large to be produced by this weak, but well understood, source.

Problems

1. A distribution of particles obeys the diffusion equation in one dimension

$$\frac{\partial f(x, t)}{\partial t} = \frac{\partial}{\partial x}\left(D(x)\frac{\partial f}{\partial x}\right)$$

At $x = 0$ a source injects particles so that the net current in the x direction is S_0. At x_1 a sink absorbs all particles reaching x_1:

(a) If $D(x) = D_0 = a$ constant, find the steady-state distribution $f(x)$.
(b) If $D(x) = D_0 e^{ax}$ find the steady-state distribution.

2. (a) The radial diffusion equation for equatorially trapped particles is

$$\frac{\partial f(\mu, J, L)}{\partial t} = \frac{\partial}{\partial L}\left[\frac{D_{LL}}{L^2}\frac{\partial}{\partial L}(L^2 f)\right]$$

Let $D_{LL} = D_0 L^{10}$ and assume that there is a source of particles at $L = 7$. All particles are absorbed by the atmosphere at $L = 1$. If $f(L = 7) = f_0$, find the steady-state L dependence of the distribution function. Show that the net current is diffusing inward.

(b) With the same conditions as above, except that $D_{LL} = D_0 L^n$, find the L dependence of $f(L)$.

(c) If $n = 1$ for the conditions given in part (b) find $f(L)$. Sketch $f(L)$.

3. The diffusion equation in a two-dimensional cartesian coordinate system is

$$\frac{\partial f(x, y, t)}{\partial t} = \frac{\partial}{\partial x}D\frac{\partial f}{\partial x} + \frac{\partial}{\partial y}D\frac{\partial f}{\partial y}$$

Let $f = X(x)Y(y)T(t)$, separate the variables and write the general solution for the equation where $D = D_0 = $ constant and

$$f = 0 \text{ at } x = 0 \text{ and at } y = 0$$
$$f = 0 \text{ on the lines } x = x_1 \text{ and } y = y_1$$

4. Consider the diffusion of particles in an infinite cylindrical medium. Assume that there is no z or ϕ dependence of the distribution. The diffusion equation is

$$\frac{\partial f(r, t)}{\partial t} = \nabla \cdot D\nabla f$$

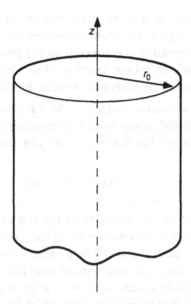

Assume $D = D_0$ = constant. Write the diffusion equation in cylindrical co-ordinates. Separate the time and spatial variables and find the lowest-order eigenfunction and eigenvalue. Assume that all particles are absorbed at the surface $r = r_0$.

5. A group of 5 kev protons with $\alpha_{eq} = 90°$ is injected at $L = 7$ in a dipole field. If they diffuse with μ = constant and J = constant, find the energy of the protons when they reach $L = 5$, $L = 3$ and $L = 1.5$.

6. The radial diffusion equation is given below where f is the number of particles per unit L, per unit μ = magnetic moment, and per unit J = integral invariant:

$$\frac{\partial f}{\partial t} = \frac{\partial}{\partial L}\left[\frac{D_{LL}}{L^2}\frac{\partial}{\partial L}(L^2 f)\right]$$

Assume that the plasma sheet is injecting a steady net electron source S_0 (electrons per unit $\Delta\mu$ per unit ΔJ) into the radiation belt at $L = 7$ and assume that there are no pitch-angle scattering losses. Also assume that D_{LL} is constant ($D_{LL} = D_0$) everywhere except in the interval $3 < L < 4$ where D_{LL} is infinite:

(a) Find the expression for f in the interval $1 < L < 3$.
(b) Find the expression for f in the interval $3 < L < 4$.
(c) What is the value of f at $L = 7$?

Answers should be in terms of D_0, S_0 and L.

7. One source of high-energy protons in the radiation belt is the decay of neutrons which are produced in the atmosphere by collisions of cosmic rays with oxygen and nitrogen nuclei. Each neutron then decays into a proton, an

electron, and a neutrino, and if this decay occurs while the neutron is moving through the trapping region, the electron and proton may become trapped. In the decay process momentum is conserved so the proton moves with approximately the velocity of the parent neutron. The electron is emitted with an energy of several hundred kilovolts in a random direction:

(a) The half-life of a neutron (which is the time interval in which half the neutrons in an initial group decay) is approximately 10 minutes. Therefore, the probability that a neutron will not have decayed by time t is given by

$$P(t) = \exp(-\lambda t)$$

Show that $\lambda = 1.16 \times 10^{-3}$ s^{-1}.

(b) If a 98 MeV neutron is emitted from the top of the atmosphere on the equator in the zenith direction, what is the probability that the neutron will decay inside the $L = 5$ shell? What will be the J value of the trapped proton? (Neglect the relativistic time dilation factor.)

(c) If the neutron decay occurs at $L = 2$, what is the probability that the resulting electron will be in a trapped orbit? (Assume a dipole field and neglect the atmosphere.)

8. Neglecting the phase coordinates, three independent coordinates are needed to specify a trapped particle's trajectory, the most fundamental being the three adiabatic invariants μ, J, and Φ. Which of the following sets are suitable to completely define a trapped particle's trajectory in a dipole field? Assume that you know the particle species:

$$(\cos \alpha_{eq}, E, L), (B_{eq}, B_m, \alpha_{eq})$$
$$(B_{eq}, \cos \alpha_{eq}, p), (B_{eq}/B_m, E, L)$$
$$(E, p, L), (L, J, p), (L, \Phi, \mu)$$
$$(\alpha_{eq}, B_m, \Phi), (L, J, \tau_b), (E, \mu, \Phi)$$

where

α_{eq} = equatorial pitch angle
B_{eq} = equatorial magnetic field of guiding center path
B_m = magnetic field at mirror point
p = scalar momentum
E = kinetic energy

7

Diffusion in pitch angle

Diffusion in the pitch angles of trapped particles is an important re-distribution and loss mechanism. Observations of particles with pitch angles inside the loss cone indicate that this process takes place at all L values in which trapping occurs, although the process proceeds more rapidly with increasing L value. At $L \approx 6$, which is the magnetic shell whose field lines connect to the auroral zone, rapid pitch-angle diffusion of electrons is a common occurrence. Electrons are fed into the loss cone by multiple deflections, and their subsequent motion into the atmosphere in both the northern and southern hemispheres supplies energy to the polar aurora.

Electron diffusion by collisions with atmospheric atoms

Collisions of electrons with atmospheric atoms is one cause of pitch-angle diffusion. While collisions are the dominant loss for electrons at only very low L values ($L < 1.3$), they occur at all L for those electrons which mirror at low altitudes. It is a well-understood process, and for this reason it is instructive to derive the diffusion coefficient from the basic formula describing the scattering of electrons by atoms. Because of their greater mass, protons and heavier ions are not scattered appreciably in pitch angle by collisions. The cumulative effect of collisions on ions is to reduce the ion velocity to thermal values while leaving their direction largely unchanged. The following treatment of electron collisions with the atmosphere is appropriate for regions in which the ambient atmosphere has only a small effect during the electron drift period. The diffusion coefficients will therefore be obtained by averaging over a complete drift cycle.

Convenient coordinates for pitch-angle diffusion calculations are E, L and $x = \cos \alpha_{eq}$. The value of L is not altered by a collision, as the guiding center can move at most only two gyroradii in a single collision, and the change in L is usually much less. The electron energy will be almost unchanged when the electron is deflected by the much heavier nucleus of an atom. The loss of electron energy by collisions with bound or free electrons is important and will be included later by adding a term to the standard Fokker–Planck equation. Thus, for the pitch-angle diffusion process x is the only coordinate involved and $\langle (\Delta x)^2 \rangle / 2$ is the only diffusion coefficient to be evaluated. The equatorial pitch angle, or some other function of α_{eq}, could equally well be used as the variable, but the equations are somewhat simpler using the cosine of the pitch angle. The local pitch angle, or some function of it, would not be a useful variable as it is not constant during the unperturbed bounce motion of the electron.

The cross-section for scattering of electrons by the nucleus of a neutral atom is

$$\sigma(\eta) = \frac{z^2 e^4}{64 \pi^2 \varepsilon_0^2 m_0^2 c^4} \frac{1 - \beta^2}{\beta^4} \frac{1}{\sin^4 \dfrac{\eta}{2}} \tag{7.1}$$

where z is the atomic number of the atom, ε_0 is the electric permitivity of free space and η is the scattering angle. Because of the $\sin^4 \eta/2$ term in the denominator the cross-section is large for small deflections, and this feature justifies the assumption inherent in the Fokker–Planck equation that the deflection in an individual interaction is small. To calculate $\langle (\Delta x)^2 \rangle$ we will first find the average change per unit time in the local pitch-angle cosine due to collisions. With this collision average we will compute the equivalent change in the cosine of the equatorial pitch angle. Then this change will be averaged over the bounce motion of the particle, weighted at each portion of the path by the density of scattering centers at that path increment and by the time the particle spends in that increment. Finally, the average over longitude will be computed giving the diffusion coefficient $\langle (\Delta x)^2 \rangle / 2$.

The geometry of an electron scattering event is illustrated in Figure 7.1, where $\mathbf{v}$ and $\mathbf{v}'$ are the electron velocities before and after the collision. If the local pitch angles before and after the collision are α and α', respectively, and the angle through which the electron is scattered is η, the change in $\cos \alpha$ produced by a particular collision is (see Figure 7.1)

$$\begin{aligned}
\Delta \cos \alpha &= \cos \alpha' - \cos \alpha \\
&= \cos \alpha \cos \eta + \sin \alpha \sin \eta \cos \psi - \cos \alpha \\
&= -2 \cos \alpha \sin^2 \frac{\eta}{2} + (1 - \cos^2 \alpha)^{1/2} \sin \eta \cos \psi
\end{aligned} \tag{7.2}$$

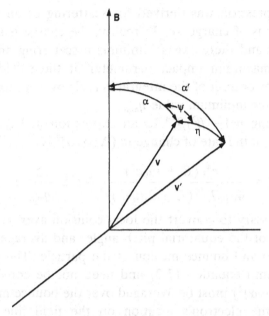

Figure 7.1. Geometry of particle scattering through angle η which changes the velocity from $\mathbf{v}$ to $\mathbf{v}'$, and changes the pitch angle from α to α'.

The average time rate of change in $\cos\alpha$ is the value of $\Delta\cos\alpha$ from equation (7.2) multiplied by the probability of that change occurring in unit time and integrated over all possible $\Delta\alpha$. The probability per unit time of an electron scattering through angle η into unit solid angle is $Nv\sigma(\eta)$, where N is the density of scattering centers. If the sum over all scattering possibilities is denoted by the use of curly brackets, the time rate of change of any quantity $g(\eta, \psi)$ by collisions is given by

$$\{g\} = \int_0^{2\pi} d\psi \int_0^{\pi} g(\eta, \psi) Nv\sigma(\eta)\sin\eta\, d\eta \qquad (7.3)$$

The collision average time rate of change in $(\Delta\cos\alpha)^2$ is obtained by squaring $\Delta\cos\alpha$ in (7.2) and substituting $(\Delta\cos\alpha)^2$ for $g(\eta, \psi)$ in (7.3). The cross-term in the square of (7.2) contains $\cos\psi$ and vanishes after integrating over ψ. The two remaining terms can be compared for relative value by noting that $\sin\eta\, d\eta = 4\sin(\eta/2)\, d(\sin(\eta/2))$.

The square of the first term of (7.2) when inserted in (7.3) has a η dependence of $\int_0^{\pi} \sin(\eta/2)\, d(\sin(\eta/2)$, which is of order 1. The square of the second term contains a divergent factor $\int_0^{\pi} d(\sin\eta/2)/(\sin\eta/2)$ and will therefore be the dominant term. The logarithmic divergence results from the coulomb cross-section (7.1), which becomes infinite for $\eta = 0$. The cross-section expression was derived for scattering of an electron by an unshielded nucleus of charge ze. In reality, the charge is shielded by the orbital electrons and there exists a minimum scattering angle η_{min} corresponding to a maximum impact parameter at the shielding boundary. Therefore, the lower limit of the integral in (7.3) over η should not be zero but should be some minimum value η_{min}.

After converting $v(1 - \beta^2)\beta^{-4}$ to an expression in kinetic energy the collision average of the rate of change in $(\Delta \cos \alpha)^2$ is

$$\{(\Delta \cos \alpha)^2\} = \frac{e^4 c(E + m_0 c^2)}{4\pi\varepsilon_0^2 E^{3/2}(E + 2m_0 c^2)^{3/2}} Nz^2 \ln \frac{2}{\eta_{min}}(1 - \cos^2 \alpha) \quad (7.4)$$

It is now necessary to convert the local collision average to the change in x, the cosine of the equatorial pitch angle, and average these changes over the gyration and bounce motion of the particle. The gyration phase does not enter into equation (7.2) and need not be considered further. However, $\{(\Delta \cos \alpha)^2\}$ must be averaged over the bounce motion as α and N depend on the electron's location on the field line. The relation between $\Delta \cos \alpha$ and $\Delta \cos \alpha_{eq} = \Delta x$ is obtained from the fact that the magnetic moment is constant along the bounce:

$$1 - \cos^2 \alpha = (1 - x^2)\left(\frac{B(s)}{B_{eq}}\right) \quad (7.5)$$

Differentiating (7.5) leads to

$$\cos \alpha \, d(\cos \alpha) = x \, dx \left(\frac{B(s)}{B_{eq}}\right) = x \, dx \frac{(1 - \cos^2 \alpha)}{(1 - x^2)}$$

and

$$\{(\Delta x)^2\} = \{(\Delta \cos \alpha)^2\}(dx/d\cos \alpha)^2$$

$$= \{(\Delta \cos \alpha)^2\}\frac{\cos^2 \alpha}{x^2}\frac{(1 - x^2)^2}{(1 - \cos^2 \alpha)^2} \quad (7.6)$$

The bounce average is obtained by integrating (7.6) between mirror points weighting each element of the path by the time a particle spends in that increment:

$$\langle(\Delta x)^2\rangle = \int_{s_m}^{s'_m}\{(\Delta x)^2\}\frac{ds}{v \cos \alpha} \bigg/ \int_{s_m}^{s'_m}\frac{ds}{v \cos \alpha} \quad (7.7)$$

$$\langle(\Delta x)^2\rangle = \frac{e^4 c(E + m_0 c^2)}{4\pi\varepsilon_0^2 E^{3/2}(E + 2m_0 c^2)^{3/2}}\int_{s_m}^{s'_m}\frac{ds}{v \cos \alpha}\frac{\cos^2 \alpha}{x^2}$$

$$\times \frac{(1 - x^2)^2}{(1 - \cos^2 \alpha)} N(s) z^2 \ln \frac{2}{\eta_{min}} \times \frac{1}{\int_{s_m}^{s'_m} \frac{ds}{v \cos \alpha}} \qquad (7.8)$$

If the atmosphere contains a mixture of elements, each with a different z_i, N_i and $\eta_{i,min}$, then $N(s) z^2 \ln (2/\eta_{min})$ in equation (7.8) is replaced by $\sum_i N_i z_i^2 \ln (2/\eta_{i,min})$. Because $N(s)$ increases rapidly as s increases and the electron samples the lower atmosphere, $\langle (\Delta x)^2 \rangle$ will be a strong function of x. In general, equation (7.8) must be computed numerically using an atmospheric model for $N(s)$.

Finally, the value of $\langle (\Delta x)^2 \rangle$ must be averaged over the drift in longitude. If the geomagnetic field were a centered dipole, this averaging would not be necessary. However, there is considerable distortion of the field at low altitudes where the atmosphere is important to trapped particles. Therefore, in longitude or drift averaging, one must take account of the fact that the particle 'sees' a different atmosphere at each longitude, and the northern and southern halves of its bounce trajectory pass through different air densities. Furthermore, the angular drift velocity varies with longitude. These factors are usually accounted for by constructing an 'average' atmosphere based on the atmospheric density along traces of constant B, L about the Earth, the density at each point being weighted inversely with drift rate. This average atmosphere is then used to evaluate the integral of equation (7.8).

The energy loss which electrons experience in collisions with free and bound electrons can be included in equation (6.33) by adding an additional term to the Fokker–Planck equation

$$\left. \frac{\partial f}{\partial t} \right|_{energy\ loss} = -\frac{\partial}{\partial E} \langle \Delta E \rangle f \qquad (7.9)$$

where $\langle \Delta E \rangle$ is the time rate of energy loss by the electron as it collides with free electrons and with the orbital electrons of neutral atoms. The loss of energy as an electron penetrates material, the so-called dE/dx (in this expression dx is the differential of length and not $\cos \alpha_{eq}$) is well known. The average time rate of energy loss is therefore

$$\{\Delta E\} = v \cdot \frac{dE}{dx} = -\frac{e^4}{4\pi \varepsilon_0^2 m_0 c \beta} \sum_i z_i N_i \ln \frac{E(E/m_0 c^2 + 2)^{1/2}}{I_i} \qquad (7.10)$$

where I_i is the mean ionization potential for an atom of the ith species. Again it is necessary to perform bounce and drift averages of the atmosphere to obtain an 'average' atmosphere which the trapped electron will experience. The trajectory average of $\{\Delta E\}$ is then

$$\langle \Delta E \rangle = \frac{1}{\tau_d} \int_0^{2\pi} \frac{\mathrm{d}\phi}{\dot{\phi}} \cdot \frac{2}{\tau_b} \int_{s_m}^{s'_m} \frac{\{\Delta E(s, \phi)\}\,\mathrm{d}s}{v \cos \alpha(s)} \tag{7.11}$$

where $\dot{\phi}$ is the longitudinal drift rate (a function of ϕ), and $\{\Delta E\}$ from equation (7.10) is a function of longitude and latitude through the densities N_i.

In the coordinates E, x and L the Fokker–Planck equation, including the energy loss term, is (from 6.33)

$$\frac{\partial f(E, x, L)}{\partial t} = -\frac{\partial}{\partial E} \langle \Delta E \rangle f$$

$$+ \frac{\partial}{\partial x}\left[\frac{\langle (\Delta x)^2 \rangle}{2} x N_2(x) \frac{\partial}{\partial x}\left(\frac{f}{x N_2(x)}\right)\right] \tag{7.12}$$

where $N_2(x)$ is defined by equation (6.31).

In equation (7.12) the pitch angle and energy variables cannot be separated to allow an eigenfunction solution. The effect of energy loss is to mix the normal modes of the pitch-angle distribution so that they do not decay independently. Hence, an initial distribution in a single mode will evolve into several pitch-angle modes as time passes. This behavior prevents a simple solution by the separation of variables, although approximate solutions by this method have been useful.

Equation (7.12) can be evaluated by finite difference techniques. A straightforward application was the computation of the evolution over time of electrons injected into the magnetosphere by the Starfish nuclear weapon effects test in 1962. Intense fluxes of electrons produced by the beta decay of fission fragments were distributed between $L = 1.12$ and $L = 7$, although the major portion was confined below $L = 2$. Since these electrons were of higher energy than most of the electrons of natural origin, and the fluxes were more intense, it was possible to measure the intensity and distribution of bomb produced electrons for many months. This experiment thus offered a unique opportunity to compare the observed loss rate of trapped electrons with the value expected from atmospheric scattering. Some of the results are shown in Figure 7.2(*a*, *b*). Figure 7.3 compares calculated and observed values of the decay time, the time required for the flux to be reduced by the factor $1/e$. In general, agreement is excellent below $L = 1.3$, but above that limit electrons are removed much more rapidly than the theory permits. This result indicates that some other process, such as scattering by electromagnetic waves, is responsible for the observed diffusion in pitch angle.

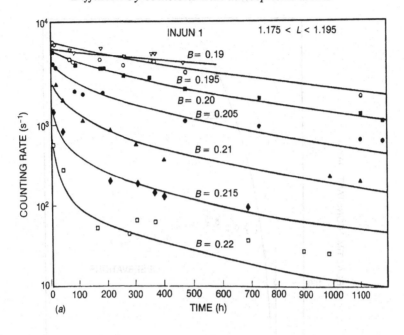

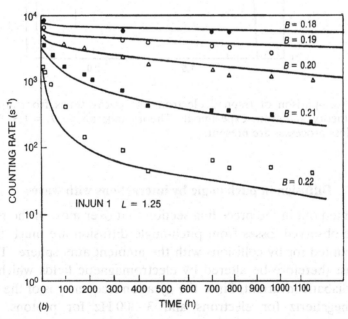

Figure 7.2(*a*) Loss of Starfish electrons at $L = 1.185$ by scattering with the ambient atmosphere. Symbols show the experimental values of omnidirectional flux of electrons (> 1 MeV), and lines are fluxes obtained from numerical integration of equation (7.12). (*b*) Same as Figure 7.2(*a*) but for $L = 1.25$. (From *J. Geophys. Res.* (1964) **69**, 397.)

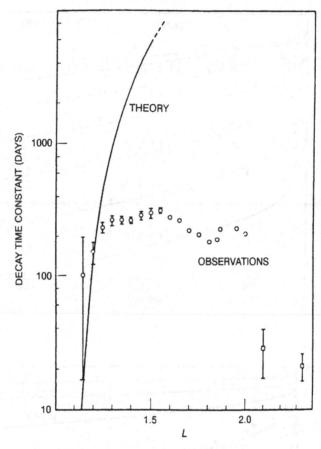

Figure 7.3. Comparison of trapped electron flux decay times from atmospheric scattering theory and from experiment. Theory fails above $L = 1.3$, indicating that other loss processes are present.

Diffusion in pitch angle by interactions with waves

It was pointed out in the preceding section that over most of the magneto-sphere the observed losses from pitch-angle diffusion are much too large to be accounted for by collisions with the ambient atmosphere. The pitch angles must therefore be altered by electromagnetic fields which change the first adiabatic invariant. Since the gyration frequency is of the order of 5 kHz–1 megahertz for electrons and 3–300 Hz for protons, electro-magnetic field variations at these or higher frequencies are required to alter the first adiabatic invariant and thereby change the pitch angle.

Many types of plasma waves occur in the magnetosphere. From the standpoint of trapped radiation, circularly polarized whistler and ion cyclotron waves appear to be the most important for their effects on trapped electrons and ions respectively. Since electron interactions with the whistler mode waves have received the most attention and are the easiest to calculate, they will be described here. The case for protons and ion cyclotron waves is similar except that the velocity of the protons is comparable to the phase velocity of the waves, and some of the approximations introduced for the electron case are not valid.

A whistler or right-hand circularly polarized wave propagating parallel to the geomagnetic field will have **E** and **b** wave vectors perpendicular to the magnetic field. The situation is depicted in Figure 7.4, which indicates the sense of rotation of the vectors. For a wave of this type the phase velocity is given by a dispersion relation which relates the phase velocity to the frequency:

$$v_{ph} = \frac{c[w(\Omega_e - w)]^{1/2}}{\omega_p} \tag{7.13}$$

where $\omega_p = (e^2 N / \varepsilon_0 m_e)^{1/2}$ is the plasma frequency of the medium, ω is the wave frequency, Ω_e is the electron gyration frequency, N is the electron

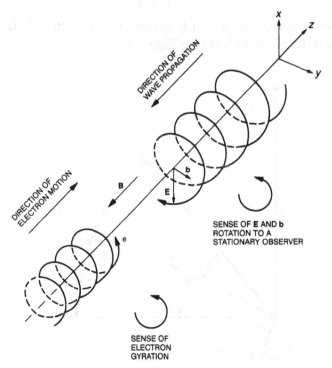

Figure 7.4. Interaction of a gyrating electron with a right-hand circular polarized wave propagating parallel to the magnetic field.

number density and m_e is the electron mass. For the phase velocity to be real, $\omega < \Omega_e$. The amplitudes of the **b** and **E** vectors are related by

$$\frac{|\mathbf{E}|}{|\mathbf{b}|} = v_{\text{ph}}$$

The phase velocity depends on the magnetic field intensity through Ω_e and on the ambient electron density through ω_p. Phase velocities of whistler waves are usually less than 0.1c. Therefore, for energetic electron interactions $v_{\text{ph}}/v \ll 1$.

In the stationary frame of reference the wave magnetic field is

$$\mathbf{b} = b[\hat{\mathbf{e}}_x \cos(\omega t + kz) - \hat{\mathbf{e}}_y \sin(\omega t + kz)] \tag{7.14}$$

for a wave moving in the negative z direction. For an electron whose guiding center moves in the positive z direction at velocity v_z:

$$z = v_z t + z_0 \tag{7.15}$$

The electron will therefore experience the Doppler shifted wave as

$$\mathbf{b} = b\{\hat{\mathbf{e}}_x \cos[(\omega + kv_z)t + kz_0] - \hat{\mathbf{e}}_y \sin[(\omega + kv_z)t + kz_0]\} \tag{7.16}$$

where the Doppler shifted frequency is

$$\omega_d = \omega + kv_z$$

The electron will see the **E** and **b** vectors rotate with angular frequency $\omega + kv_z$, and the phase of **b** is (see Figure 7.5).

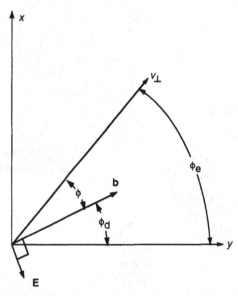

Figure 7.5. Definition of ϕ, the phase angle between the wave **b** vector and $\mathbf{v}_\perp$, the electron perpendicular velocity.

$$\phi_d = \int \omega_d \, dt = \int (\omega + k v_z) \, dt \tag{7.17}$$

The perpendicular velocity vector of the electron, $\mathbf{v}_\perp$, gyrates about the field line with phase

$$\phi_e = \int \Omega_e \, dt + \phi_0 = \Omega_e t + \phi_0$$

Thus, the angular difference between $\mathbf{v}_\perp$ and $\mathbf{b}$ is

$$\phi = \phi_e - \phi_d = (\Omega_e - \omega - k v_z) t + \phi_0 \tag{7.18}$$

where ϕ_0 is the initial phase difference between ϕ_e and ϕ_d at time $t = 0$.

The electric and magnetic components of the wave exert forces on the electron which, in the non-relativistic case, are, for $q = -e$,

$$\dot{\mathbf{v}} = -\frac{e}{m} [\mathbf{E} + \mathbf{v} \times \mathbf{b}] \tag{7.19}$$

The components of $\dot{\mathbf{v}}$ perpendicular and parallel to z are

$$\dot{v}_z = -\frac{e}{m} b v_\perp \sin \phi \tag{7.20}$$

$$\dot{v}_\perp = \frac{e}{m} \{ E \sin \phi + b v_z \sin \phi \} = \frac{e}{m} b (v_{ph} + v_z) \sin \phi \tag{7.21}$$

where use has been made of $|\mathbf{E}|/|\mathbf{b}| = v_{ph}$. The change in v_z and $v_\perp$ will cause changes in the electron kinetic energy and pitch angle. The rate of energy change is given by

$$\frac{d}{dt} \left(\frac{m}{2} (v_z^2 + v_\perp^2) \right) = m(v_z \dot{v}_z + v_\perp \dot{v}_\perp)$$

$$= -v_z e b v_\perp \sin \phi + v_\perp e b (v_{ph} + v_z) \sin \phi$$

$$= e b v_{ph} v_\perp \sin \phi \tag{7.22}$$

Only the electric field term from equation (7.21) contributes to (7.22), as is expected since deflection by the magnetic field will not change the particle energy.

The pitch angle α changes at a rate

$$\dot{\alpha} = \frac{d}{dt} \left[\tan^{-1} \left(\frac{v_\perp}{v_z} \right) \right]$$

$$= \frac{v_z \dot{v}_\perp - \dot{v}_z v_\perp}{v_z^2 + v_\perp^2} \tag{7.23}$$

Using (7.20) and (7.21), equation (7.23) becomes

$$\dot{\alpha} = \frac{e}{m} b \left(1 + \frac{v_{ph} \cos \alpha}{v} \right) \sin \phi \tag{7.24}$$

In general, ϕ is a rapidly changing function of time given by equation (7.18). In this case $\dot{\alpha}$ will display a rapid sinusoidal variation, and the accumulated deflection will not be large. However, if the parallel velocity v_z of the particle has the resonant value

$$v_z = \frac{\Omega_e - \omega}{k} \qquad (7.25)$$

then ϕ from (7.18) and thus $\dot{\alpha}$ from (7.24) will be constant, and appreciable changes in α can accumulate. From equation (7.22), changes in energy will also take place if ϕ is constant over an appreciable time interval. The electron is said to be in resonance with the wave. The resonant frequency of the wave is given by $\omega = \Omega - kv_z$.

The sign of $\dot{\alpha}$ and $\mathrm{d}(\tfrac{1}{2}mv^2)/\mathrm{d}t$ is determined by the phase angle ϕ. If $0 < \phi < \pi$ both α and the energy increase. For these values of ϕ, $\mathbf{v}_\perp \cdot \mathbf{E} < 0$ (see Figure 7.5), and the electron would be expected to gain energy at the expense of energy in the wave. If $\mathbf{v}_\perp \cdot \mathbf{E} > 0$ ($\pi < \phi < 2\pi$), the electron will lose energy and decrease its pitch angle. In a flux of electrons uniformly distributed in the azimuthal angle, ϕ_e of Figure 7.5, some electrons will be deflected to smaller α and some to larger α during encounters with waves of finite length. The overall effect of many such encounters will be a diffusion in pitch angle.

A single frequency wave of infinite length will impart net deflections only for particles in exact resonance. Waves even slightly off-resonance will successively increase and decrease α as the phase angle ϕ rotates through 2π. If, however, the wave is of finite length, off-resonant frequencies can change α permanently. In general, the shorter the duration of the wave, the further off-resonance the wave can be and still contribute to the deflection.

An estimate of the width $\Delta\omega$ of the wave band contributing to $\dot{\alpha}$ for a wave of duration Δt is obtained as follows. If the change in ϕ is limited to π in time Δt:

$$\Delta\phi = (\Omega_e - \omega - kv_z)\Delta t = \pi \qquad (7.26)$$

Consider the first factor on the right as a function of ω and expand it in a Taylor series about the resonant frequency $\omega = \Omega_e - kv_z$. The derivative of k with respect to ω is the reciprocal of the wave group velocity. Thus,

$$\Delta\phi = \left\{ (\Delta\phi)_{\omega = \Omega_e - kv_z} + \frac{\partial}{\partial\omega}(\Delta\phi) \cdot \Delta\omega + \ldots \right\} = \pi$$

$$\sim -\left(1 + \frac{v_z}{v_g}\right)\Delta\omega\Delta t = \pi \qquad (7.27)$$

and

$$|\Delta t| \approx \frac{\pi}{\left(1 + \dfrac{v_z}{v_g}\right)\Delta\omega} \tag{7.28}$$

The diffusion coefficient can be estimated for a series of waves of duration Δt interacting with the particle. With brackets denoting change per unit time:

$$D_{\alpha\alpha} = \frac{\langle(\Delta\alpha)^2\rangle}{2} \approx \frac{1}{2}\left\langle\left(\frac{d\alpha}{dt}\right)^2(\Delta t)^2\right\rangle$$

$$\approx \frac{1}{2}\left(\frac{e}{m}\right)^2\frac{b^2}{\Delta\omega}\left(1 + \frac{v_{ph}\cos\alpha}{v}\right)^2(\sin^2\phi)_{ave}\frac{\pi}{\left(1 + \dfrac{v_z}{v_g}\right)}$$

$$\approx \frac{\pi}{4}\left(\frac{e}{m}\right)^2\frac{b^2}{\Delta\omega}\left(1 + \frac{v_{ph}\cos\alpha}{v}\right)^2\frac{1}{\left(1 + \dfrac{v_z}{v_g}\right)} \tag{7.29}$$

since for particles uniformly distributed in ϕ_0, $(\sin^2\phi)_{ave} = \frac{1}{2}$. The factor $b^2/\Delta\omega$ is interpreted as the power spectral density of the waves at the resonant frequency. In situations where $v_{ph} \ll v$ and $v_g \ll v_z$, $D_{\alpha\alpha}$ can be approximated as

$$D_{\alpha\alpha} \approx \frac{\pi}{4}\left(\frac{e}{m}\right)^2\left(\frac{b^2}{\Delta\omega}\right)\frac{v_g}{v_z} \tag{7.30}$$

Other approximations, also based on heuristic arguments, give slightly different results.

A more quantitative expression for the pitch-angle diffusion coefficient can be derived by expressing the wave in general form and averaging over the stochastic variations of the wave field. The magnetic field as experienced by the electron is

$$\left.\begin{array}{l}\mathbf{b} = b_x(t)\hat{\mathbf{e}}_x + b_y(t)\hat{\mathbf{e}}_y \\ \mathbf{v}_\perp = v_\perp\cos(\Omega t + \eta)\hat{\mathbf{e}}_x + v_\perp\sin(\Omega t + \eta)\hat{\mathbf{e}}_y\end{array}\right\} \tag{7.31}$$

Returning to equation (7.24) and recognizing that $\sin\phi = |\mathbf{v}_\perp \times \mathbf{b}|/v_\perp b$,

$$\dot\alpha = \frac{e}{m}\left(1 - \frac{v_{ph}\cos\alpha}{v}\right)\left[\frac{v_x b_y - v_y b_x}{v_\perp}\right]$$

$$= K[\cos(\Omega t + \eta)\cdot b_y(t) - \sin(\Omega t + \eta)b_x(t)] \tag{7.32}$$

where K is equal to the first two factors of equation (7.32).

To find $\Delta\alpha$ equation (7.32) is integrated over a finite time interval containing several gyrations. In fact, the integration time will be long compared to the coherence time of the wave. Under these conditions the two terms on the right-hand side will contribute equally to $\Delta\alpha$ and only one need be calculated:

$$(\Delta\alpha)^2 = 4K^2 \int_0^t \cos(\Omega\xi' + \eta) b(\xi') \,d\xi' \int_0^t \cos(\Omega\xi'' + \eta) b(\xi'') \,d\xi'' \qquad (7.33)$$

Expanding the cosine terms, averaging over the electron initial phase angle η, and rearranging the integrands gives

$$(\Delta\alpha)^2 = 4K^2 \int_0^t d\xi' \int_0^t d\xi'' b(\xi') b(\xi'') \tfrac{1}{2} \cos \Omega(\xi'' - \xi') \qquad (7.34)$$

Let $\xi'' - \xi' = \tau$,

$$(\Delta\alpha)^2 = 2K^2 \int_0^t d\xi' \int_{-\xi'}^{t-\xi'} d\tau \, b(\xi') b(\xi' + \tau) \cos \Omega\tau \qquad (7.35)$$

Now exchange the order of integration and modify the limits as needed:

$$(\Delta\alpha)^2 = 2K^2 \left\{ \int_{-t}^0 d\tau \cos \Omega\tau \int_{-\tau}^t b(\xi') b(\xi' + \tau) \,d\xi' \right.$$
$$\left. + \int_0^t d\tau \cos \Omega\tau \int_0^{t-\tau} b(\xi') b(\xi' + \tau) \,d\xi' \right\} \qquad (7.36)$$

The integral $1/t \int_0^t b(\xi') b(\xi' + \tau) \,d\xi'$ is the auto-correlation function of a component of the wave magnetic field. It is usually written as $\langle b(\xi') b(\xi' + \tau) \rangle$. These correlation integrals are functions only of the 'lag' τ and are small for lags larger than the correlation length or coherence time of the wave.

Equation (7.36) then becomes, replacing τ by $-\tau$ in the first term,

$$(\Delta\alpha)^2 = 2K^2 \left\{ \int_0^t d\tau \cos \Omega\tau (t - \tau) \langle b(\xi') b(\xi' - \tau) \rangle \right.$$
$$\left. + \int_0^t d\tau \cos \Omega\tau (t - \tau) \langle b(\xi') b(\xi' - \tau) \rangle \right\} \qquad (7.37)$$

Because $\langle b(\xi') b(\xi' + \tau) \rangle = \langle b(\xi') b(\xi' - \tau) \rangle$ and the value is small unless τ is less than the coherence time, $(t - \tau) \approx t$. Also, because the integrand is zero at large τ, the upper limit of the integrals can be increased to infinity:

$$(\Delta\alpha)^2 = 4K^2 t \int_0^\infty d\tau \cos \Omega\tau \langle b(\xi') b(\xi' + \tau) \rangle = K^2 t P_b'(\Omega) \qquad (7.38)$$

where

$$P_b'(\Omega) = 4 \int_0^\infty d\tau \langle b(\xi') b(\xi' + \tau) \rangle \cos \Omega\tau \qquad (7.39)$$

is the power spectral density at the gyration frequency for a component of the wave as measured in the moving frame of the particle guiding center.

This spectral density is related to the power spectral density in the rest frame at the Doppler shifted frequency by

$$P'_b(\Omega) = P_b(\Omega - kv_z)\frac{d(\Omega - kv_z)}{d\Omega} = \frac{P_b(\Omega - kv_z)}{[1 + v_z/v_g]} \qquad (7.40)$$

The diffusion coefficient is then

$$D_{\alpha\alpha} = \frac{(\Delta\alpha)^2}{2t} = \frac{1}{2}\left(\frac{e}{m}\right)^2\left(1 - \frac{v_{ph}\cos\alpha}{v}\right)^2\frac{P_b(\Omega - kv_z)}{[1 + v_z/v_g]}$$
$$\approx \frac{1}{2}\left(\frac{e}{m}\right)^2\left(\frac{v_g}{v_z}\right)P_b(\Omega - kv_z) \qquad (7.41)$$

for $v \gg v_{ph}$ and $v_z \gg v_g$.

Equation (7.41) is the local value of the diffusion coefficient describing the change in the local pitch angle. To calculate the behavior of trapped particles it is necessary to convert the pitch angle to some quantity which is constant during the bounce and then average over the bounce motion. If one chooses $\cos\alpha_e$ as the variable, the procedure to be followed is given in equations (7.6) and (7.7).

The energy loss described in equation (7.22) indicates that the electrons change energy as they diffuse in pitch angle. The magnitude of this energy change for a given $\Delta\alpha$ can be estimated from equations (7.22) and (7.24):

$$\frac{\Delta E}{E} = \frac{1}{E}\frac{dE}{d\alpha}\Delta\alpha = \frac{1}{E}\frac{dE}{dt}\frac{dt}{d\alpha}\Delta\alpha$$
$$= \frac{2mv_{ph}v_\perp v}{mv^2(v + v_{ph}\cos\alpha)}\Delta\alpha$$
$$\sim \frac{2v_{ph}v_\perp}{v^2} \qquad (7.42)$$

for $v_{ph} \ll v$ and for $\Delta\alpha = 1$. If $v_{ph} \ll v$, the fractional change in energy will be small. Therefore, for electrons it is usually permissible to ignore the energy change during pitch-angle diffusion and use a diffusion equation with some function of the pitch angle as the independent variable. With independent variables L, E and x the only diffusion term is the one containing D_{xx} and the equation to be used is (6.33).

A more graphic description of the diffusion in pitch angle can be obtained by transforming to a frame of reference moving with the phase velocity of the wave. Since $\mathbf{v}_{ph} = (\mathbf{E} \times \mathbf{b})/b^2$, the electric field of the wave will be zero in that frame:

$$\mathbf{E}' = \mathbf{E} + \mathbf{v}_{ph} \times (\mathbf{B}_0 + \mathbf{b}) = \mathbf{E} + \mathbf{v}_{ph} \times \mathbf{b} = 0 \qquad (7.43)$$

and the electron energy will be conserved. This condition expressed in terms of $\mathbf{v}_\perp$ and $\mathbf{v}_\| = \mathbf{v}_z$ is

$$\tfrac{1}{2}d[v_\perp^2 + (v_\| - v_{ph})^2] = v_\perp\,dv_\perp + (v_\| - v_{ph})\,dv_\| = 0 \qquad (7.44)$$

This differential equation in velocity $(v_\perp, v_\|)$ space describes an element of a circle whose center is located at $-v_{ph}$ on the $v_\|$ axis (see Figure 7.6). This diffusion path differs from the constant energy path centered at the origin, although the difference will be small if $v_{ph} \ll v$. As the particle moves along the diffusion path of equation (7.44) the parallel velocity will change and the resonant frequency will also change. The motion along the line will take place in a number of small steps, each increment being in a random direction and resulting from interaction with a wave.

From equation (7.22) it is clear that an individual electron will either lose or gain energy depending on the azimuthal phase angle ϕ. However, the overall flow of the diffusing electrons depends on the distribution of the particles in velocity space. Figure 7.7 illustrates a hypothetical distribution function $F(v_\|, v_\perp)$, where $F(v_\|, v_\perp)2\pi v_\perp\,dv_\perp\,dv_\|$ is the number of particles within the differential element of velocity space. In this diagram more particles are at large pitch angles, and the net diffusion is towards smaller α and lower energies.

The general form of the diffusion equation for wave–particle interactions can be derived by assuming that the current of particles flowing along the diffusion path of equation (7.44) is proportional to the slope of F along that path. The change in F along an increment $d\mathbf{v}$ is

$$\mathbf{dv} \cdot \nabla F = \frac{\partial F}{\partial v_\perp}\,dv_\perp + \frac{\partial F}{\partial v_\|}\,dv_\| \qquad (7.45)$$

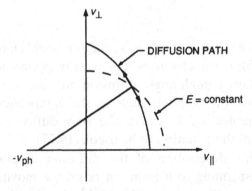

Figure 7.6. Diffusion path of a particle interacting with waves compared to the constant energy path. As the pitch angle decreases, the particle energy decreases.

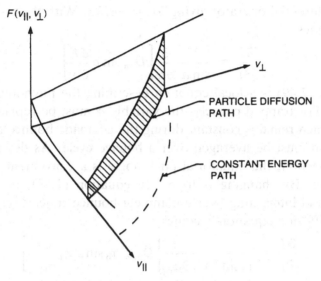

Figure 7.7. Particle distribution in velocity space showing the diffusion path and the constant energy path.

But $dv_\perp$ and $dv_\parallel$ are related by (7.44), which, combined with (7.45), yields

$$\text{current} \propto -D\left(v_\perp \frac{\partial F}{\partial v_\parallel} - (v_\parallel - v_{ph})\frac{\partial F}{\partial v_\perp}\right) = -D\mathcal{D}_v F \qquad (7.46)$$

The proportionality constant D is recognized as the diffusion coefficient, and the differential operator $\mathcal{D}_v$ computes the gradient of F along the diffusion path. The rate of change in F is equal to the divergence of the current and is given by

$$\frac{\partial F}{\partial t} = \frac{1}{\mathcal{J}}\mathcal{D}_v[D\mathcal{J}\mathcal{D}_v F] \qquad (7.47)$$

where $\mathcal{J}$ is the Jacobian ($\mathcal{J} = v_\perp$), relating a volume element in velocity space $dv_x \, dv_y \, dv_z$ to $v_\perp \, d_\phi \, dv_\perp \, dv_\parallel$. More rigorous methods have confirmed the validity of (7.47).

To reduce (7.47) to a more familiar form for a simplified case assume that $v_\parallel \gg v_{ph}$ so that only the pitch angle α changes during diffusion. The change from $(v_\perp, v_\parallel)$ to (v, α) coordinates is obtained by inserting

$$\left.\begin{aligned}
\frac{\partial}{\partial v_\perp} &= \sin\alpha\frac{\partial}{\partial v} + \frac{\cos\alpha}{v}\frac{\partial}{\partial\alpha} \\[2mm]
\frac{\partial}{\partial v_\parallel} &= \cos\alpha\frac{\partial}{\partial v} - \frac{\sin\alpha}{v}\frac{\partial}{\partial\alpha}
\end{aligned}\right\} \qquad (7.48)$$

into the differential operator giving $\mathcal{D}_v = -\partial/\partial\alpha$. With these substitutions (7.47) becomes

$$\frac{\partial F}{\partial t} = \frac{1}{\sin\alpha}\frac{\partial}{\partial\alpha}\left[D_{\alpha\alpha}\sin\alpha\frac{\partial F}{\partial\alpha}\right] \tag{7.49}$$

Equation (7.49) is a local equation describing the pitch-angle diffusion at a point. For trapped particles the variable α must be replaced by some quantity which remains constant during the adiabatic bounce motion, and the equation must be averaged over a bounce cycle. As described in the preceding section, the equatorial pitch angle is a convenient variable for this purpose. By changing α to α_{eq} in equation (7.49), multiplying by $ds/v\cos\alpha$ and integrating over a complete bounce trajectory, the bounce averaged diffusion equation becomes

$$\frac{\partial F}{\partial t} = \frac{1}{\tau_b\sin 2\alpha_{eq}}\frac{\partial}{\partial\alpha_{eq}}\left[\bar{D}_{\alpha_{eq}\alpha_{eq}}\tau_b\sin 2\alpha_{eq}\frac{\partial F}{\partial\alpha_{eq}}\right] \tag{7.50}$$

where the averaged diffusion coefficient is

$$\bar{D}_{\alpha_{eq}\alpha_{eq}} = \frac{1}{\tau_b}\oint D_{\alpha_{eq}\alpha_{eq}}(s)\frac{ds}{v\cos\alpha} \tag{7.51}$$

This equation is entirely equivalent to (6.33). By changing the distribution function $f(x, E, L)$ in (6.33) to a velocity or phase space distribution function using (6.32), and by replacing the independent variable $x = \cos\alpha_{eq}$ by α_{eq}, equation (6.33) becomes (7.50).

Coupling of particle and wave energy

The diffusion in pitch angle by waves which are not produced by the particles themselves is sometimes termed parasitic diffusion. If the power spectrum of the waves is known, the change in the particle distribution can be calculated using equation (7.50) for the diffusion equation and (7.51) with (7.41) for the diffusion coefficient. This approach is satisfactory for estimating the diffusion rates from waves which are not produced by the particles themselves.

As described in the derivation of equation (7.22), particles can exchange energy with the waves, either gaining or losing energy depending on the phase angle between $\mathbf{v}_\perp$ and $\mathbf{b}$. The wave is augmented or reduced by the electric and magnetic fields produced by the particle currents. The wave response is characterized by a growth rate γ where

$$\frac{d}{dt}|b|^2 = 2\gamma|b|^2 \tag{7.52}$$

A positive γ denotes a growing wave. The form of γ is given by

$$\gamma(\omega) \approx -g(\omega) \int_0^\infty dv_\perp v_\perp^2 \left[v_\perp \frac{\partial F}{\partial v_\parallel} - (v_\parallel - v_{ph}) \frac{\partial F}{\partial v_\perp} \right]_{v_\parallel = \Omega_e - \omega/k} \tag{7.53}$$

where $g(\omega)$ is a slowly varying function of ω. The integrand of (7.53) contains the differential operator of (7.46) and represents the slope of the distribution function along the diffusion path.

With reference to Figure 7.8 the integrand is $v_\perp^2$ times the derivative of F along the diffusion path. The integral over $v_\perp$ then includes the contributions of these derivatives at all $v_\parallel = (\Omega_e - \omega)/k$. The growth rate will be large if the slope of F along the diffusion path is large and a net flow of electrons occurs towards smaller α and lower energy.

The coupling of the particle distribution with wave growth rates changes the dynamic behavior of the wave and particle systems. If the waves propagate parallel to **B**, the conservation of wave energy is expressed by

$$\frac{\partial b^2}{\partial t} + v_g \frac{\partial b^2}{\partial z} = 2(\gamma - \ell) b^2 \tag{7.54}$$

where ℓ represents the internal loss rate of wave energy. Equations (7.41), (7.47), (7.53) and (7.54) relate the wave and particle behavior. An initial

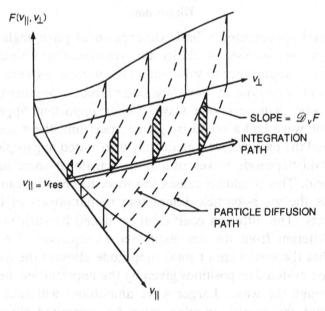

Figure 7.8. Effect of particle distribution on the growth rate of waves. The growth rate for a resonant $v_\parallel$ (equation (7.43)) contains the slope of the distribution along the diffusion path integrated over $v_\perp$.

distribution of particles with an empty loss cone will generally lead to wave growth. The growing waves will increase $D_{\alpha\alpha}$ by equation (7.41), leading to more rapid diffusion and loss of particles into the loss cone. This process should provide a regulation mechanism for trapped radiation. An increased source of electrons would lead to greater wave growth and increased loss rates, thereby limiting the particle flux.

While the concepts of a flux limiting mechanism are valid, the complex geometry and inhomogeneities in the magnetosphere make quantitative calculations ambiguous. In particular, the propagation of wave energy out of the particle interaction region reduces the effectiveness of the flux limiting mechanism. Idealized calculations usually assume that the waves travel only parallel to **B** and are reflected at each end of the field line with reflectivity $\mathcal{R}$. These assumptions lead to a steady-state situation in which

$$1/\mathcal{R} = \exp\left(\gamma\Delta s/v_{\mathrm{ph}}\right) \tag{7.55}$$

where Δs is the length of the wave–particle interaction region. Equation (7.55) simply states that the loss of wave energy at each reflection is balanced by the growth during passage through Δs.

Discussion

This chapter has presented a basic description of pitch-angle scattering. However, in the interests of clarity this treatment was simplified, and many factors of importance to wave–particle interactions were ignored. In particular, only one type of wave was considered, a parallel propagating electromagnetic, whistler-mode wave. Other important approximations were that the wave had a broad frequency spectrum, the wave amplitude was small and the particles were uniformly distributed in gyrophase.

In fact, whistler-mode waves usually propagate at some angle to the magnetic field. This condition causes the wave to be elliptically polarized and extends the wave–particle interaction to harmonics of the particle gyrofrequency. The diffusion coefficient produced by such waves is substantially different from the one expressed in equation (7.41). The assumption that the waves are of small amplitude allowed the forces on the particle to be evaluted at positions given by the unperturbed motion of the particle through the wave. Larger wave amplitudes will alter the trajectories so that the particle motion must be computed throughout the encounter. In extreme cases, the particle can become trapped in the fields of the wave and the resonance time is thus greatly extended.

Problems

1. Using equation (7.12), but neglecting the energy loss term, show that an isotropic flux will remain isotropic regardless of the form of the atmospheric collision coefficient $\langle (\Delta x)^2 \rangle$.

2. D_{xx} is the diffusion coefficient in terms of $x = \cos \alpha_{eq}$, but we wish to use B_m, the magnetic field at the mirroring point, as the independent variable. Find the expression for $\langle (\Delta B_m)^2 \rangle$ in terms of D_{xx}, B_{eq} and B_m.

3. Starting with equation (7.2) and retaining only terms of order $\ln 2/\eta_{min}$, show that the first and second Fokker–Planck coefficients for atmospheric scattering (before bounce averaging) satisfy the relationship (6.20).

4. Derive the bounce averaged pitch-angle diffusion equation (7.50) from the local equation (7.49).

5. If the reflectivity of the ionosphere for electromagnetic wave energy is 0.2 and the equatorial interaction region at $L = 2$ is 10^3 km in length, what must be the growth rate in the interaction region to sustain a 5 kHz parallel propagating wave? Assume that the cold plasma is hydrogen with a density of 2×10^9 m^{-3}.

8

Diffusion in the *L* coordinate or radial diffusion

Particle diffusion through random increments in the *L* coordinate is frequently termed radial diffusion because the process changes the radial distances of trapped particles from the Earth. This type of diffusion is crucial in forming the radiation belts as it provides a mechanism for transporting particles from the outer boundary of the magnetosphere into the inner belt. It also leads to the redistribution of particles injected or accelerated during magnetic storms and substorms. While radial diffusion may be overshadowed at times by the massive injections which occur during large storms and substorms, its role in bringing particles inward, in accelerating trapped particles and in redistributing newly injected particles is of major importance.

Since the third invariant Φ is proportional to L^{-1}, radial diffusion must proceed by fluctuations in the third invariant. Variations in a trapped particle's third invariant require changes in the electric or magnetic fields that are more rapid than the particle drift frequency. Drift periods vary from tens of seconds to about a day (see Appendix B), hence, perturbations over a wide range of frequencies can alter the third invariant. Because the gyration and bounce periods are much shorter than the drift period, the first and second invariants are less likely to be affected by many of these field perturbations.

The paths of the mirror points of particles undergoing third invariant diffusion but with constant first and second invariant are shown in Figure 8.1. The trapped particles mirroring on the equatorial plane remain on that plane, as demanded by the need to keep $J = 0$. The mirror points of particles mirroring off the equatorial plane move along lines of almost constant latitude, the latitude increasing slightly with increasing *L*. As particles diffuse inward (outward) the momentum increases (decreases) in order to maintain a constant value for the magnetic moment.

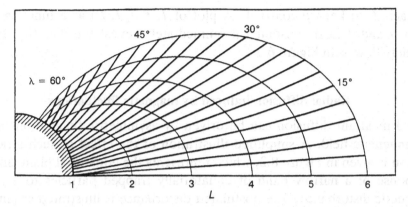

Figure 8.1. Diffusion paths of the mirroring points of trapped particles undergoing radial diffusion conserving the first and second adiabatic invariants.

In Chapter 6 a diffusion equation ((6.26)) for L shell diffusion was derived for a distribution function in the variables μ, J and L:

$$\frac{\partial f(\mu, J, L)}{\partial t} = \frac{\partial}{\partial L}\left[\frac{\langle(\Delta L)^2\rangle}{2}\frac{1}{L^2}\frac{\partial}{\partial L}(L^2 f)\right] \tag{8.1}$$

where $\langle(\Delta L)^2\rangle/2 = D_{LL}$ is the radial diffusion coefficient. In this chapter the diffusion coefficient will be evaluated for magnetic and electric perturbations, as both of these variations are common in the magnetosphere. To make the computation for D_{LL} easier, idealized models for the geometry of the perturbations will be assumed. While these simplifications are not completely justified, they illustrate the principals involved and are appropriate in view of present limited knowledge of the geometry of the magnetic and electric field variations.

It is expected that D_{LL} will depend on general statistical properties such as the power spectrum of a multitude of disturbances rather than on details of any single fluctuation. In each event particles will be moved inward or outward depending on their location at the time of the field changes. Summed over many events the cumulative motion of an individual particle may be inward or outward. However, the overall flow of particles will depend on the distribution in L of the particle populations as described in Chapter 6. If the coordinates μ, J and L are used, the flow will be inward wherever $\partial(L^2 f)/\partial L$ is positive.

The partial derivatives with respect to L in (8.1) are taken with μ and J constant. It is not possible to determine the overall direction of particle flow by plotting $j(E, \alpha_{\text{eq}} = (\pi/2))$ as a function of L. One must first convert $j(E, \alpha_{\text{eq}})$ into $f(\mu, J = 0, L)$ using equation (6.35) and selecting E

at each L to keep μ constant. A plot of $L^2 f(\mu, J, L)$ as a function of L with μ and J held constant will immediately reveal the direction of net particle flow as in Figure 6.3.

Radial diffusion induced by magnetic fluctuations

Third invariant diffusion can be driven by asymmetric fluctuations in the geomagnetic field. A simplified illustration of the effect of such a perturbation is given in Figure 8.2, which represents the equatorial plane and the response of a narrow band of equatorially trapped particles to a global magnetic disturbance. The postulated disturbance is illustrated schematically at the bottom of the diagram and consists of a sudden compression of the magnetosphere by an increase in solar wind pressure. After this compressive impulse the solar wind pressure gradually decreases, allowing the magnetosphere to relax to its original configuration. The compression will be greatest on the sunward side of the Earth (right-hand side of Figure 8.2). During this initial compression, the trapped particles are carried inward to the dashed line, this transport taking place before the particles have an opportunity to drift appreciably in longitude. Particles on the sunward side of the Earth are moved by the largest amount and are left in more intense magnetic fields. The dashed line showing the post-compression particle positions is not a line of constant B, nor is it the path of drifting particles. It is the instantaneous location of the particles which

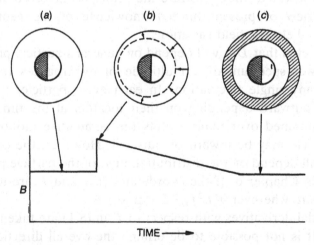

Figure 8.2(*a–c*) Effect of an asymmetric, sudden compression and slow relaxation of the geomagnetic field on a narrow band of equatorially trapped particles. After the recovery period the particles fill the shaded band.

were previously drifting in the narrow band of Figure 8.2(*a*). The sudden compression will change the values of the third invariant but not the values of μ and J. Particles on the sunward side will suffer the largest change in Φ and the largest increase in energy. Following this compression, the geomagnetic field relaxes slowly, keeping all adiabatic invariants constant. Meanwhile, each particle drifts about the Earth at constant μ, J and Φ and, after a number of orbits, the natural dispersion in drift velocity results in the broad band of particles depicted in Figure (8.2(*c*)). During this field relaxation all particles are moved outward as the magnetic field recovers. The overall effect of the sudden compression and slow relaxation is to move some particles inward (those initially on the sunward side of the Earth) and to transport some particles outward (those initially on the nightside). Many events of this type, each small in overall effect, will lead to a diffusion in the L coordinates of the particles.

The motion of the trapped particles under magnetic perturbation depends both on the magnetic field change and on the induced electric field. The induced electric field cannot be calculated directly from the magnetic field change as only $\nabla \times \mathbf{E}$ is given by $\partial \mathbf{B}/\partial t$. However, if one assumes that the cold plasma in the magnetosphere is a good electrical conductor in the direction parallel to the magnetic field and that the Earth itself, or the ionosphere, is a perfect conductor, then the induced electric field is completely specified. Where these conditions apply, the apparent velocity $\mathbf{v}_f$ of the field line and the induced electric field are related by

$$\mathbf{v}_f = (\mathbf{E} \times \mathbf{B})/B^2 \tag{8.2}$$

The instantaneous position of an individual magnetic field line is obtained by tracing its position through space, beginning with its fixed position at the conducting Earth or ionosphere. The position of the field line at subsequent time intervals permits a computation of $\mathbf{v}_f$ and, by equation (8.2), an evaluation of $\mathbf{E}$. Since equation (8.2) also describes the electric drift velocity of ions and electrons, these particles will be carried with the moving field line. This simultaneous motion of particles and magnetic field is called the frozen field condition. While the frozen field condition obtains over much of the magnetosphere, it is not universally valid. In the outer magnetosphere where the cold plasma density is low, electric fields parallel to $\mathbf{B}$ occur, and the frozen field assumption is invalid. However, in the present idealized calculation, this assumption will be followed with the understanding that in the actual magnetosphere the induced electric field may be quite different.

The general approach used to find D_{LL} is to construct an idealized model of the field disturbance and to compute the radial displacement of a trapped particle experiencing the disturbance. This displacement is then squared and averaged over all possible initial longitudes of the particle and over many disturbances occurring at random times. In this way a diffusion coefficient is obtained as a function of the statistical characteristics of the disturbances. The full derivation of D_{LL} will be performed for equatorially trapped particles and the work extended later to other pitch angles.

Equations for the guiding center drift velocity were derived in Chapter 2 (equation (2.32)). Because we are considering only equatorially trapped particles, the curvature drift term is zero. In an electric field perpendicular to an inhomogeneous magnetic field **B** the perpendicular drift velocity is

$$\mathbf{v}_\perp = -\frac{1}{qB^2}\mathbf{B} \times (q\mathbf{E} - \mu\nabla B) \tag{8.3}$$

Let the magnetic field be composed of a dipole field $\mathbf{B}_d$ and a disturbance field **b** where $\mathbf{b} \ll \mathbf{B}_d$. The fact that the disturbance is much smaller than the dipole field will allow the disturbance field to be considered a perturbation on the usual gradient **B** and electric field drifts. In this case it is assumed that the magnetic changes are caused by magnetospheric boundary currents, which then produce field changes interior to the boundary. Equation (3.23) is therefore suitable to express the disturbance field in terms of spherical harmonics:

$$\mathbf{b} = -\nabla\Psi = -\nabla R_E \sum_{n=1}^{\infty}\left(\frac{r}{R_E}\right)^n \sum_{m=0}^{n}(\bar{g}_n^m \cos m\phi + \bar{h}_n^m \sin m\phi)P_n^m(\cos\theta) \tag{8.4}$$

For small perturbations near the Earth, only the leading terms with $n < 3$ are important. A further simplification results from aligning the dipole perpendicular to the solar wind and labeling the meridian containing the Sun $\phi = 0$. Thus $\bar{h}_n^m = 0$ for all n, m and $\bar{g}_n^m = 0$ when $n + m$ is even. The only terms remaining are $\bar{g}_1^0$ and $\bar{g}_2^1$ and the disturbance field becomes

$$\begin{aligned}
\mathbf{b}(t) &= -\nabla\left[\bar{g}_1^0 r \cos\theta + \sqrt{3}\frac{r^2}{R_E}\bar{g}_2^1 \sin 2\theta \cos\phi\right] \\
&= [-S(t)\cos\theta - A(t)r\sin 2\theta\cos\phi]\hat{\mathbf{e}}_r \\
&\quad + [S(t)\sin\theta - A(t)r\cos 2\theta\cos\phi]\hat{\mathbf{e}}_\theta \\
&\quad + A(t)r\cos\theta\sin\phi\hat{\mathbf{e}}_\phi
\end{aligned} \tag{8.5}$$

By expressing **b** in the form of (8.5) the Maxwell equation $\nabla \cdot \mathbf{B} = 0$ is automatically satisfied. The time-dependent coefficients $S(t)$ and $A(t)$ are parameters representing the parts of the disturbance field which are symmetric and asymmetric respectively, about the polar axis.

The induction electric field will be computed from the motion of the field lines by

$$\mathbf{E} = -\mathbf{v}_f \times \mathbf{B} \tag{8.6}$$

where the field line velocity $\mathbf{v}_f$ is obtained by tracing the equatorial crossing of a field line whose feet are fixed at a lower altitude, taken for convenience at the origin of the dipole.

The field line equations in polar coordinates are given by

$$\frac{\mathrm{d}r}{B_r} = \frac{r\,\mathrm{d}\theta}{B_\theta}, \quad \frac{r\sin\theta\,\mathrm{d}\phi}{B_\phi} = \frac{r\,\mathrm{d}\theta}{B_\theta} \tag{8.7}$$

with $\mathbf{B} = \mathbf{B}_d + \mathbf{b}$.

In the spirit of perturbation theory these equations can be integrated from the origin to the equator by replacing r and ϕ in the expressions for B_r, B_θ and B_ϕ by their dipole values

$$r = R_0 \sin^2\theta, \quad \phi = \phi_0 \tag{8.8}$$

This approximation sets the field $\mathbf{B}_d + \mathbf{b}$ at each position of the undistorted path equal to the distorted magnetic field values. With the additional assumption that $A(t)$ and $S(t)$ are much smaller than B the equations can be integrated analytically from $\theta = 0$ to $\pi/2$ to give r and ϕ in the equatorial plane in terms of $A(t)$ and $S(t)$ and the constants R_0 and ϕ_0. These constants are the radial distance and longitude of the equatorial crossing of the undistorted field line. The changes in r and ϕ as a function of $A(t)$ and $S(t)$ can then be interpreted as motion of the field line, and the resulting electric field can be calculated from equation (8.6). The electric field in the equatorial plane obtained in this manner is

$$\mathbf{E} = \frac{1}{7}r^2\frac{\mathrm{d}A}{\mathrm{d}t}\sin\phi\hat{\mathbf{e}}_r + r\left(\frac{1}{2}\frac{\mathrm{d}S}{\mathrm{d}t} + \frac{8}{21}r\frac{\mathrm{d}A}{\mathrm{d}t}\cos\phi\right)\hat{\mathbf{e}}_\phi \tag{8.9}$$

The magnetic symmetry assumed for $\mathbf{b}$ gives $b_r = b_\phi = E_\theta = 0$ in the equatorial plane. The radial component of the drift velocity of equatorial particles from equation (8.3) reduces to

$$\frac{\mathrm{d}r}{\mathrm{d}t} = \left(-\frac{E_\phi}{B} + \frac{\mu}{qBr}\frac{\partial b_\theta}{\partial\phi}\right) \tag{8.10}$$

Substituting values for E_ϕ and b_θ from equations (8.9) and (8.5) and using the dipole value for B in (8.10) results in

$$\frac{\mathrm{d}r}{\mathrm{d}t} = -r\left(\frac{1}{2B_d}\frac{\mathrm{d}S}{\mathrm{d}t} + \frac{8}{21}\frac{r}{B_d}\frac{\mathrm{d}A}{\mathrm{d}t}\cos\phi\right) - \frac{\mu}{qB_d}A\sin\phi \tag{8.11}$$

In equation (8.11) the time-dependent quantities on the right-hand side are the coefficients A and S and the particle coordinates r and ϕ. In keeping with the usual policy of perturbation theory we will use the unperturbed values of these coordinates for the particle position. Therefore, on the right-hand side of equation (8.11) set $r = r_0$ and $\phi = \Omega_D t + \eta$ where Ω_D is the angular drift frequency of the particle and η is the particle longitude at $t = 0$. The magnetic moment can be replaced by its value in terms of the angular drift velocity $\mu = \Omega_D r_0^2 q/3$. Integrating equation (8.11) over time from zero to t gives the radial displacement at time t:

$$\int_{r_0}^{r(t)} dr = r(t) - r_0 = -\frac{5}{7} \frac{r_0^2 \Omega_D}{B_d} \int_0^t A(\xi) \sin(\Omega_D \xi + \eta) d\xi$$

$$- \frac{r_0}{2B_d}[S(t) - S(0)] - \frac{8}{21} \frac{r_0^2}{B_d}\{A(t) \cos(\Omega_D t + \eta) - A(0) \cos\eta\}$$

$$(8.12)$$

With the exception of the first term on the right-hand side, all terms are bounded and of order b/B_d. On the other hand, the integral term can grow without limit as t increases, provided $A(\xi)$ has frequencies in the neighborhood of Ω_D. This term is therefore the important one for radial displacements.

Only the asymmetric part of the disturbance field survives in computing radial displacements. This result is as expected since symmetric compressions and relaxations will return particles to their original radial positions. Also, electric and magnetic drifts contribute almost equally to the coefficient 5/7 of the dominant term. Therefore, the assumptions regarding the induced electric field are quite important to the result.

The diffusion coefficient is constructed from the average value of the square of the radial displacement. The technique is similar to the one used to derive pitch-angle diffusion coefficients in Chapter 7. The square of (8.12), keeping only the dominant term, can be manipulated to give

$$[r(t) - r_0]^2 = \left(\frac{5}{7}\right)^2 \left(\frac{r_0^2 \Omega_D}{B_d}\right)^2 \int_0^t d\xi' A(\xi') \sin(\Omega_D \xi' + \eta)$$

$$\times \int_0^t d\xi'' A(\xi'') \sin(\Omega_D \xi'' + \eta)$$

$$= \left(\frac{5}{7}\right)^2 \left(\frac{r_0^2 \Omega_D}{B_d}\right)^2 \int_0^t d\xi' \int_0^t d\xi'' A(\xi')A(\xi'')$$

$$\times \sin(\Omega_D \xi' + \eta) \sin(\Omega_D \xi'' + \eta) \qquad (8.13)$$

Equation (8.13) can be modified to bring out the physical content. Expand the sine terms using the trigonometric sums of angles formula and multiply the two factors to give

$$\sin(\Omega_D \xi' + \eta)\sin(\Omega_D \xi'' + \eta) = \sin \Omega_D \xi' \sin \Omega_D \xi'' \cos^2 \eta$$

$$+ \sin \Omega_D \xi' \cos \Omega_D \xi'' \cos \eta \sin \eta$$

$$+ \cos \Omega_D \xi' \sin \Omega_D \xi'' \sin \eta \cos \eta$$

$$+ \cos \Omega_D \xi' \cos \Omega_D \xi'' \sin^2 \eta \qquad (8.14)$$

The quantity needed for the diffusion coefficient is $(r(t) - r_0)^2/t$ averaged over initial particle positions in longitude and averaged over a representative sample of the magnetic fluctuations. Averaging (8.14) over η eliminates the two terms containing $\sin \eta \cos \eta$ and replaces the $\sin^2 \eta$ and $\cos^2 \eta$ factors by $\frac{1}{2}$. The result is

$$[r(t) - r_0]^2 = \frac{1}{2}\left(\frac{5}{7}\right)^2 \left(\frac{r_0^2 \Omega_D}{B_d}\right)^2 \int_0^t d\xi' \int_0^t d\xi'' A(\xi')A(\xi'') \cos \Omega_D(\xi'' - \xi')$$

$$(8.15)$$

In averaging over η it was assumed that the particles were evenly distributed over η or drift phase. Since the magnetic disturbance is asymmetric, the particle distribution after the compression will not be uniform. Any subsequent disturbance would then act on a non-uniform phase distribution, and the $\sin \eta \cos \eta$ terms in (8.14) would not be zero. However, because the angular drift velocity depends on particle energy and pitch angle, in time the dispersion in Ω_D will restore the uniform distribution in η. This assumption of an efficient phase mixing is usually made in derivations of D_{LL}.

Now change the inner variable of integration to $\zeta = \xi'' - \xi'$ where ζ varies from $-\xi'$ to $t - \xi'$ giving

$$[r(t) - r_0]^2 = \frac{1}{2}\left(\frac{5}{7}\right)^2 \left(\frac{r_0^2 \Omega_D}{B_d}\right)^2 \int_0^t d\xi' \int_{\xi'}^{t-\xi'} d\zeta A(\xi')A(\xi' + \zeta) \cos \Omega_D \zeta \quad (8.16)$$

$A(\xi')$ is assumed to fluctuate randomly with zero mean. Over a sufficiently long period of time integrals such as

$$\frac{1}{t}\int_0^t A(\xi') A(\xi' + \zeta) d\xi' = \frac{1}{t}\int_0^t A(\xi')A(\xi' - \zeta) d\xi' \qquad (8.17)$$

will be equal and will be independent of the time interval chosen. They depend only on the 'lag', ζ, which is the difference in the arguments of the two factors in the integrand. Now, reverse the order of integration in the double integral of (8.16) and use (8.17) to simplify the result:

$$[r(t) - r_0]^2 = \left(\frac{5}{7}\right)^2 \left(\frac{r_0^2 \Omega_D}{B_d}\right)^2 \int_0^t d\zeta \cos \Omega_D \zeta \int_0^t d\xi' \, A(\xi') A(\xi' + \zeta) \quad (8.18)$$

The inner integral is t times the auto-correlation function of $A(\xi')$, which is written $\langle A(\xi')\,A(\xi' + \zeta)\rangle$. It is a function of ζ, not of ξ', and its value will be large when ζ is small so that $A(\xi')$ and $A(\xi' + \zeta)$ are nearly equal. For large ζ, $A(\xi')$ and $A(\xi' + \zeta)$ are uncorrelated and are as likely as not to have different signs. For ζ greater than this correlation length, the autocorrelation function will be zero. As long as the time interval t is larger than the correlation period, the integration over ζ can be extended to infinity giving

$$[r(t) - r_0]^2 = \left(\frac{5}{7}\right)^2 \left(\frac{r_0^2 \Omega_D}{B_d}\right)^2 t \int_0^\infty d\,\zeta \langle A(\xi')\,A(\xi' + \zeta)\rangle \cos \Omega_D \zeta \quad (8.19)$$

The diffusion coefficient in terms of $(\Delta r)^2$ is for magnetic field fluctuations

$$D_{LL}^M = \frac{\langle(\Delta L)^2\rangle}{2} = \frac{[r(t) - r_0]^2}{R_E^2 2t} = \frac{1}{8}\left(\frac{5}{7}\right)^2 \left(\frac{r_0^2 \Omega_D}{R_E B_d}\right)^2 P_A(\Omega_D) \quad (8.20)$$

where

$$P_A(\Omega_D) = 4\int_0^\infty d\,\zeta \langle A(\xi')A(\xi' + \zeta)\rangle \cos \Omega_D \xi \quad (8.21)$$

is the power spectral density of the field variation evaluated at the drift frequency. Thus, the radial diffusion coefficient will be large when the magnetic fluctuations occur at frequencies near the particle drift frequency.

The diffusion coefficient can be expressed in more familiar terms by setting $r_0 = LR_E$, $\Omega_D = 2\pi v_{\text{drift}}$, and $B_d = B_0/L^3$. With these substitutions

$$D_{LL}^M = \frac{\pi^2}{2}\left(\frac{5}{7}\right)^2 \frac{R_E^2 L^{10}}{B_0^2} v_{\text{drift}}^2 P_A(v_{\text{drift}}) \quad (8.22)$$

The variables in equation (8.22) are the L value and drift frequency which is a function of L and μ. For non-relativistic particles $v_{\text{drift}} \propto \mu/L^2$ so that D_{LL}^M is influenced by the v dependence of $P_A(v)$. In the special case where $P_A(v)$ varies as v^{-2}, D_{LL}^M will have no v_{drift} dependence, and particles of all energies will diffuse at the same rates. If the power spectrum varies as v^{-n}, D_{LL}^M will be proportional to $L^{6+2n}\mu^{2-n}$. Since the magnitude of D_{LL}^M depends directly on $P_A(v_{\text{drift}})$ and the L variation depends on the spectral content, it is to be expected that observed values of D_{LL}^M and their L dependence will change with global magnetic activity.

A similar calculation for off-equatorial particles is more complex but follows the same principals. The curvature drift term must be included in equation (8.3) and the projected change in r at the equator must be averaged over the complete bounce motion, weighting the contribution at each field line segment by the time the particle spends in that segment. The result of this averaging is the mirroring latitude correction factor,

$\Gamma(\alpha_{eq})$, shown in Figure 8.3. This factor is the ratio of the diffusion coefficient at pitch angle α_{eq} to the diffusion coefficient at $\alpha_{eq} = \pi/2$. The magnetic perturbations are most effective in diffusing particles with large equatorial pitch angles so that diffusion proceeds most rapidly for particles confined to the equatorial plane.

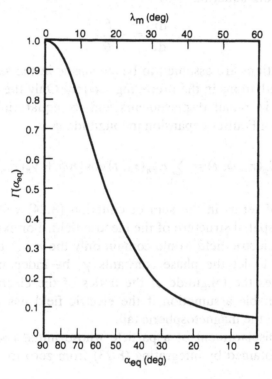

Figure 8.3. Latitude-dependent factor of the radial diffusion coefficient for magnetic fluctuations. The curve shows the more rapid diffusion of equatorially trapped particles.

Radial diffusion induced by electric potential fields

Large-scale electric potential fields are imposed on the magnetosphere by the solar wind and by plasma circulation within the magnetosphere. However, the magnitude and geometry of these fields is uncertain at present, and estimates of diffusion from this mechanism are somewhat speculative. Nevertheless, it is important to estimate the magnitude and character of diffusion from electric potential fields in order to assess the importance of this mechanism.

The calculation of electric field diffusion proceeds in much the same way as diffusion by magnetic perturbations. Again, the development will be restricted initially to equatorial particles with the results for off-equatorial particles considered at the end. The applied electric field will be assumed perpendicular to $\mathbf{B}_d$ at all positions in the magnetosphere. The starting point for the calculation is the radial component of the $\mathbf{E} \times \mathbf{B}/B^2$ drift velocity from equation (8.3)

$$\frac{\mathrm{d}r}{\mathrm{d}t} = -\frac{E_\phi}{B_d} \tag{8.23}$$

The time variations are assumed to be stochastic in the same sense as the magnetic perturbations in the preceding section. Only the ϕ component of $\mathbf{E}$ is involved in radial displacements and its equatorial values can be represented by a Fourier expansion in longitude ϕ:

$$E_\phi(r_0, \phi, t) = \sum_{n=1}^{N} E_{\phi n}(r_0, t) \cos\left[n\phi + \gamma_n(r_0, t)\right] \tag{8.24}$$

The number of terms in the sum of equation (8.24) will depend on the complexity or spatial structure of the electric field. For example a uniform dawn-to-dusk electric field would contain only the $n = 1$ term. A simplification will be to let the phase constants γ_n be independent of t. This assumption fixes the longitude of the nodes of the electric field components, a reasonable assumption if the electric field has its origin in the solar wind or in the magnetospheric tail.

The radial displacement of a particle whose initial coordinates are r_0 and $\phi = \eta$ is obtained by integrating (8.23) from zero to t replacing ϕ by $\Omega_D t + \eta$:

$$r(t) - r_0 = -\frac{1}{B_d} \int_0^t \sum E_{\phi n}(r_0, \xi) \cos\left[n\Omega_D \xi + n\eta + \gamma_n(r_0)\right] \mathrm{d}\xi \tag{8.25}$$

With this equation the expression for $\langle (\Delta r)^2 \rangle$ is obtained in the same manner as for magnetic perturbations. Equation (8.25) is squared and averaged over η. The averaging over initial longitude η will eliminate all terms except the power spectrum expressions. Also, only the fluctuating part of the electric field has an influence on the motion. This result is expected. A steady electric field will distort the azimuthal drift path, but the orbit will remain closed and no net displacement will occur. Squaring equation (8.25), averaging over longitude, and rearranging the integrals as in equation (8.12)–(8.22) results in the diffusion coefficient

$$D_{LL}^E = \left\langle \frac{(\Delta L)^2}{2} \right\rangle = \frac{1}{2R_E^2 B_d^2} \sum_{n=1}^{N} \int_0^t \langle \widetilde{E}_{\phi n}(r_0, \xi') \widetilde{E}_{\phi n}(r_0, \xi' + \zeta) \rangle \cos \Omega_D \zeta \, d\zeta$$

(8.26)

where the fluctuating part of the electric field is denoted by coefficients $\widetilde{E}_{\phi n}$.

For particles mirroring off the equatorial plane the diffusion coefficient can be derived by starting with equation (8.23) with L replacing r as the radial coordinate. If the electric field is always perpendicular to $\mathbf{B}_d$, the instantaneous change in L for a particle at latitude λ is given by

$$R_E \frac{dL}{dt} = -\frac{E_\phi(L, \lambda)}{B(L, \lambda)} \cdot \frac{\sqrt{(1 + 3\sin^2 \lambda)}}{\cos^3 \lambda}$$

(8.27)

The first factor is the electric field drift velocity at λ and the second is the ratio of the field line separation at the equator to the field line separation at λ. This factor is needed because at λ a smaller displacement perpendicular to $\mathbf{B}_d$ is needed to traverse a given ΔL. The disturbance electric field and the dipole magnetic field map from λ to the equator as

$$E_\phi(L, \lambda) = E_\phi(L, 0)/\cos^3 \lambda$$

(8.28)

and

$$B(L, \lambda) = B(L, 0)\sqrt{(1 + 3\sin^2 \lambda)}/\cos^6 \lambda$$

Therefore the latitude factors in equation (8.27) cancel leaving

$$\frac{dL}{dt} = -\frac{1}{R_E} \frac{E_\phi(L, 0)}{B(L, 0)}$$

(8.29)

which is the same as equation (8.23) which was written for the equatorial plane. This surprising result indicates that radial diffusion by potential electric fields proceeds at the same rate for off-equatorial particles as it does for particles trapped on the equator if the field lines are equipotentials. Expressed in terms of the power spectra of the Fourier components of the electric field, the diffusion coefficient for electric fields from (8.26) is

$$D_{LL}^E(L, v_{\text{drift}}) = \frac{L^6}{8R_E^2 B_0^2} \sum_{n=1}^{N} P_n(L, nv)_{v = v\text{drift}}$$

(8.30)

where $P_n(L, nv)$ is the power spectral density of the nth harmonic of the electric field fluctuations evaluated at the same harmonic of the drift frequency. The need for harmonics stems from the fact that if the disturbance field has n nodes, it must vary at n times the particle drift frequency to maintain the resonance condition.

Because v_{drift} for constant μ depends on L, the overall variation of D_{LL}^E with L will depend on the frequency dependence of $P_n(L,v)$ as well as on the L^6 term. For example, if $P_n \propto L^0 v^{-m}$ then $D_{LL}^E \propto L^{6+2m}/\mu^m$. As in the case of magnetically driven diffusion, the process is much more rapid at larger L values.

Observed and derived values of D_{LL}

Changes in the radial distribution of trapped particles have been inter-preted as evidence for radial diffusion. Efforts to explain these observa-tions by applying equation (8.1) have led to a number of experimental determinations of D_{LL}. Two general approaches are used. If the distribu-tion in L is evolving with time, equation (8.1) is solved as an initial value problem. The observed initial distribution is specified and numerical integration of (8.1) predicts the distribution at latter times. D_{LL} is adjusted to cause the calculated distributions to match the observed ones. Values of D_{LL} have also been obtained with (8.1) by adding source and loss terms and solving for the equilibrium distribution. The boundary conditions needed are that $f(\mu, J, L)$ is equal to the experimental values at some outer boundary and falls to zero at the inner boundary $L = 1$. Again D_{LL} is varied to give a best fit. In this latter technique it is necessary to know the particle sources and losses. Except for the neutron decay source of protons, the internal sources can usually be ignored, but the loss rates from pitch-angle scattering are important.

In four instances narrow bands of electrons were injected into the magnetosphere by high-altitude nuclear weapon detonations. The subse-quent spreading of these sharp initial distributions allows a straightforward extraction of D_{LL} from experiment. The values of D_{LL} are not sensitive to the assumed loss processes, but are, of course, characteristic of diffusion only during the time immediately following the injection.

Figure 8.4 is a compilation of theoretical (dashed lines) and experi-mental (solid lines) values of the radial diffusion coefficient. As was expected from the theoretical expressions derived earlier, D_{LL} increases with L, varying as L^6 to L^{10}. This general agreement confirms that radial diffusion processes occur as described. However, improved precision in the measurements is needed and it is necessary to understand how D_{LL} responds to changing magnetic activity. There are very large differences in the coefficients obtained by these methods, indicating that the experi-mental uncertainties are large or (more likely) that the observed diffusion coefficient is time dependent. In view of the many approximations used in

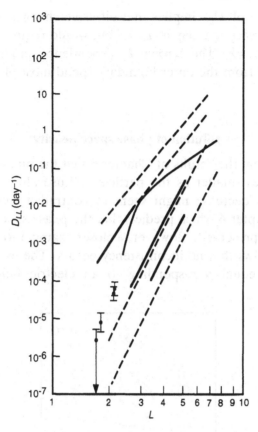

Figure 8.4. Experimental (solid lines) and theoretical (dashed lines) values of D_{LL}.

deriving theoretical values of D_{LL}, it is not surprising that theory and experiment differ somewhat. It is also expected that D_{LL} would reflect the intensity of magnetic disturbances, and these are known to vary greatly with time.

From the standpoint of theory, the assumption of small disturbance fields is quite restrictive. The phase or longitude averaging is also suspect if disturbances occur so frequently that the distribution is unable to relax to a uniform distribution in longitude before the next impulse occurs. Estimates of the particle diffusion in larger field changes and arbitrary time variations is best done by simulation, tracking a number of particles through the time-dependent electric and magnetic fields and tabulating their behavior. Again, the applicability of such results to the magnetosphere is dependent on the accuracy of the assumed field variations.

The magnitude of D_{LL} implies that diffusive changes in the L distribution could take place in a day at $L \approx 5$ but would require many days to be noticeable at $L = 2$. The strong L dependence implies that particles diffusing inward from the outer boundary spend most of their time at low L values.

Dilution of phase space density

It is apparent from the results of Chapter 6 that the phase space density of particles becomes smaller as the particles diffuse along a radial path to smaller L. This decrease might seem to contradict Liouville's theorem discussed in Chapter 4 which predicts that the phase space density along a dynamic path is preserved. However, a closer examination of the details of the diffusion shows that no inconsistency occurs. The evolution of a band of particles of equal μ responding to an electric field disturbance is

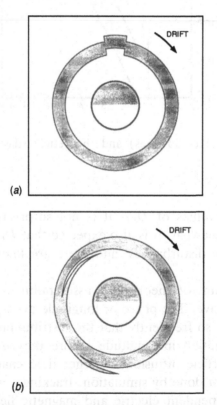

Figure 8.5. Phase space mixing after radial perturbation. Differential drift rates mix regions containing particles and voids, thereby diluting the phase space density.

illustrated in Figure 8.5. For simplicity, only equatorial particles ($J = 0$) are considered, and the disturbance is taken to be an azimuthal electric potential field in the midnight sector which acts momentarily and displaces particles in that sector outward. (The fringing field will displace particles inward at other longitudes, but this smaller motion will be ignored.) The distribution after this impulse is shown in Figure 8.5(a); all particles initially have the same μ and J which are preserved. Section (b) of the diagram illustrates the evolution of the distribution with time after the electric field is removed. The lower drift rate of particles at larger L produces a spiral in particle density, and, as time increases, the spiral becomes more tightly wound. the resulting distribution is then a fine-grained mixture of regions containing particles at densities given by Liouville's theorem and of voids containing no particles.

Eventually, the structure will become too detailed to observe, and the dispersion in drift rates for particles with slightly different energies or pitch angles will effectively mix the two regions. The overall effect is to produce a distribution in which the phase space density appears to decrease.

Problems

1. If two closely spaced field lines lie in the same meridian but are separated by Δr_{eq} at the equator and by Δr_λ at latitude λ, show that $\Delta r_{eq}/\Delta r_\lambda = \sqrt{(1 + 3\sin^2 \lambda)}/\cos^3 \lambda$.

2. An electric field is perpendicular to **B** and is in the ϕ direction throughout the magnetosphere. Show that if $E_\phi(L, \lambda)$ is its value at latitude λ, it will map to an intensity $E_\phi(L, \lambda = 0) = E_\phi(L, \lambda)\cos^3 \lambda$ at the equator.

3. It is sometimes convenient to use a distribution function proportional to the phase space density of particles, but it is also desirable to work with the radial diffusion equation in the L coordinate. Show that the diffusion equation in these terms is

$$\partial F/\partial t = L^2 \partial/\partial L (D_{LL} L^{-2} \partial F/\partial L)$$

where the partial derivatives are performed with μ and J constant.

4. Assume that a flux of equatorially trapped protons has an exponential energy spectrum $j(E) = C\exp(-E/E_0)$ and is trapped at L_1. A large-scale electric field carries the group of protons to L_2 while conserving the first and second adiabatic invariants and the phase space density. Show that the new flux also has an exponential energy spectrum with an e-folding constant of $E_0(L_1/L_2)^3$.

5. Equation (8.11) expresses dr/dt with the first term giving the contribution to dr/dt from the induction electric field and the second term giving the contribution from the magnetic field changes. Starting with (8.11) derive (8.12). What fraction of the only important term (the first) in (8.12) comes from the induction electric field and what fraction is derived from magnetic field variations?

6. Show that the angular drift frequency of an equatorially trapped particle at r_0 is given by $\Omega_D = 3\mu/qr_0^2$.

9

Summary and comments

From the principles described in the preceding chapters it is apparent that the presence of geomagnetically trapped radiation is inevitable, given the Earth's magnetic geometry, the sources of plasma in the Earth's atmosphere and solar wind, and the neutron albedo. That the outer planets of the solar system, those bodies with large magnetic dipole moments, also support radiation belts is additional evidence for the ubiquity of the particle trapping phenomena. However, the particular nature of a planet's radiation zones, the intensities, the spatial distributions, the types of particles and the energy spectra, will depend on characteristics unique to each planet. Individual planetary properties such as rotation rates, atmospheric compositions, distances from the sun, intrinsic magnetic moments, electric and magnetic fluctuations, and characteristics of plasma waves will influence the actual radiation belt that a planet can maintain.

In this regard a comparison of Earth's radiation belts with those of other planets is instructive. The solar system has a wonderful variety of radiation trapping planets whose different properties lead to unique features in their radiation belts. Venus and Mars have little or no intrinsic magnetic fields and therefore no radiation belts. Mercury has a small magnetic moment, but the distortion of its field by the solar wind precludes long-term trapping. The remaining planets, with the exception of Pluto, for which no information is available, all have intense radiation belts. The variety of conditions among the planets is impressive. The average solar wind dynamic pressure is reduced by a factor of 900 between Earth and Neptune. Magnetic moments range from 8×10^{22} A m^2 for Earth to 1.5×10^{27} A m^2 for Jupiter. The alignment of the magnetic field axis with the spin axis is almost perfect for Saturn, but these axes differ by 60° for Uranus and by 47° for Neptune. The angle between the planetary spin axis and the ecliptic plane also varies from planet to planet, being 23° at Earth,

3° at Jupiter and 98° on Uranus. All of the giant outer planets have moons and rings imbedded within their trapping zones. These bodies serve as absorbers for trapped particles, creating gaps or depressions in the fluxes. In some cases, the most notable being the satellite Io at Jupiter, the natural satellites are sources of atoms which become ionized and accelerated to form trapped particle populations. In Jupiter the density of the trapped plasma is large enough to distort the magnetic field substantially, distending the field lines in the equatorial plane.

Table 9.1 includes a list of planetary parameters pertinent to charged particle trapping. The magnetic field appears to be the crucial property since all planets with a surface magnetic field greater than 10^{-5} T have radiation belts.

As expected from these varied planetary conditions the trapped particle populations in each of the outer planets have their own characteristics. Some processes such as radial diffusion appear to proceed by quite different mechanisms on the outer planets than they do on Earth. The evidence for this belief is the observation that at Jupiter D_{LL} varies as L^4 rather than L^6 to L^{10} as on Earth. The plasma wave fields in the magnetospheres of the outer planets are also found to be significantly different from those near Earth. However, in spite of the widely different circumstances, each planet which has the appropriate magnetic geometry maintains a robust radiation belt. This fact suggests that planetary radiation belts are easily formed and do not require any special circumstances.

In the case of Earth a permanent radiation belt will be produced by the mechanisms described in the preceding chapters. If a source of plasma exists at some outer boundary, say $L = 8$–10, radial diffusion will populate the region inside this boundary. Very high-energy protons will also be introduced by the decay of cosmic ray albedo neutrons. The origin of heavier ions such as oxygen can be generally understood as derived from the atmosphere by the up-welling of ionospheric ions along polar field lines. Once these oxygen and helium ions reach the outer magnetosphere they can be trapped and subjected to radial diffusion and acceleration along with the solar wind plasma.

The characteristics of the radiation belt which would be produced by these processes can be calculated from the diffusion equations derived in Chapter 7. As a first approximation let pitch-angle diffusion remove particles at some constant rate, a reasonable assumption if pitch-angle diffusion takes place more rapidly than L-shell diffusion. Under these conditions the pitch-angle distribution is nearly in the lowest normal mode and the loss rate of electrons at a given L shell will be proportional to

Table 9.1. *Planetary characteristics*

Planet	Distance from sun (AU)	Magnetic field on surface at equator (T)	Rotation period (days)	Angle between spin axis and magnetic field axis (deg)	Obliquity angle between equatorial and orbital planes (deg)
Mercury	0.39	3.5×10^{-7}	58.6	10.0	~0
Venus	0.72	—	244.3	—	177.0
Earth	1.00	3.1×10^{-5}	1.00	11.5	23.5
Mars	1.52	—	1.03	—	24.0
Jupiter	5.20	4.1×10^{-4}	0.41	10.	3.1
Saturn	9.54	4.0×10^{-5}	0.44	<1.	26.4
Uranus	19.19	$\sim 2.3 \times 10^{-5}$	0.72	60.	97.9
Neptune	30.06	$\sim 1.3 \times 10^{-5}$	0.67	55.	28.8

the number of electrons in that shell. The approximate equation for $f(\mu, J, L, t)$, the number of particles per unit $\Delta\mu$, ΔJ and ΔL at time t is then

$$\frac{\partial f}{\partial t} = \frac{\partial}{\partial L}\left[\frac{D_{LL}}{L^2}\frac{\partial f}{\partial L}\right] - \frac{f}{\tau} + Q(\mu, J, L) \qquad (9.1)$$

where $Q(\mu, J, L)$ is a source term representing injection by neutron decay. The boundary conditions appropriate for this equation are that $f = 0$ at the inner boundary, $L = 1$, and that at an outer boundary, say $L = 8$, $f(\mu, J, L = 8)$ is equal to some constant value $f_0(\mu, J)$. The particle lifetime from pitch-angle diffusion is represented by τ, which can be a function of μ and L. For the steady state, $\partial f/\partial t = 0$, equation (9.1) can be solved numerically for arbitrary values of the functions D_{LL} and τ. The two boundary conditions are sufficient to determine the solution of this second-order differential equation. With judicious choices of D_{LL}, τ and the flux at the outer boundary this approach leads to a realistic radiation belt exhibiting many of the characteristics of the Earth's trapped particles.

With $Q = 0$ the solution to equation (9.1) for $f(\mu, J, L)$ at constant μ will be a monotonically increasing function of L. This result follows from the fact that the current of particles diffusing 'inward' is equal to $(D_{LL}/L^2)(\partial f/\partial L)$ so that $\partial f/\partial L$ is always positive. If Q is not 0, for example if particles are injected by neutron decay, f can have a maximum at some interior value of L.

However, even with $Q = 0$, the particle flux at constant energy (the quantity which is usually measured in space experiments) may show a maximum at some interior L value. For a constant μ, the particle energy must increase as the particle diffuses to smaller L and into higher B. Thus, a measurement of flux as a function of L at constant energy involves measuring particles at a different μ for each L value, and the radial profile of flux at constant energy can be quite different from the profile of f at constant μ. As mentioned before, solutions of equation (9.1) with values of $f(L = 8)$, D_{LL} and τ selected to fit experimental data produce radiation belts in fair agreement with observations. However, since the parameters describing the important processes were chosen from radiation belt data, this overall agreement is not remarkable. Although D_{LL} and τ can be estimated from theory, as was done in Chapters 7 and 8, the lack of knowledge of the electric and magnetic fields driving both pitch-angle and radial diffusion makes these values quite uncertain.

The model based on these considerations is a quiescent, steady-state radiation belt. Time variations can occur if D_{LL} and τ or the position of the outer boundary depend on variable external forces such as the solar wind velocity or density, or on the direction of the interplanetary magnetic field. However, in the intense part of the radiation belts, say $L = 1.5$–4.0, changes in particle flux would be slow, occurring only as rapidly as diffusion processes could propagate the effects of boundary conditions into the interior L shell regions.

However, observations (see Figures 5.20 and 5.21) clearly show that sudden, major disruptions in trapped particle fluxes are common and that these changes take place during magnetic storms and substorms. Magnetic substorms occur at irregular intervals of about an hour and are believed to result from a release of magnetic energy stored in the distended magneto-spheric tail. In any case the geomagnetic field becomes less distorted and more like a dipole during the expansive, or violent, phase of a substorm. This rapid (few minutes) reconfiguration of the geomagnetic field pro-duces an induction electric field in the dawn-to-dusk direction on the night side which will cause trapped particles to drift across L shells. Qualita-tively, trapped particles will drift inward and gain energy in order to keep μ constant as they drift into stronger B. The acceleration follows from the fact that the curvature and gradient drift motion of protons are in the direction of the electric field, while the electron drift is in the opposite direction. Thus, one result of a magnetic substorm is to accelerate trapped particles on the night side of the Earth and to reduce their L values. These changes occur in time intervals typically less than the drift period. Since the change in the L coordinate of a particle during a single drift period is frequently large ($\Delta L \cong 2$), this mechanism cannot be described in terms of radial diffusion, as discussed in Chapter 8. Whether the particle motion resembles diffusion or convection (or injection) depends also on the particle energy. Higher-energy particles with drift periods much less than the substorm duration (10–30 minutes) can be described by diffusion theory, while low-energy particles which experience a large ΔL during a fraction of their drift orbit must be treated by more elaborate numerical techniques.

Large magnetic storms are initiated by a sudden increase in the density or velocity of the solar plasma impacting the sunward side of the magnetosphere. The increased pressure compresses the dayside magnetosphere and in large storms can move the magnetopause inward below $5R_E$. The sudden compression of plasma at the magnetopause can produce a magnetohydrodynamic shock wave which propagates inward and can be detected on the Earth by magnetometers as a sudden increase in the geomagnetic field.

The storm compression of the magnetosphere is more pronounced on the day side and induces a westward electric field inside the magnetosphere. As in the case of substorm disturbances, this electric field will move both electrons and ions inward and increase their energy. The speed and asymmetry of the compression will change the third invariant, moving particles inward and substantially changing their distribution in the belts. In the time interval following storms or substorms the normal radial and pitch-angle diffusion processes will modify the particle distributions, smoothing out peaks and valleys and reducing gradients introduced by the sudden field changes. The normal longitudinal drift velocity varies with energy and pitch angle, and in time this drift motion will equalize the longitudinal asymmetries in the particle distributions caused by the asymmetric disturbance. Over a long-term average, the storm and substorm events behave as impulsive sources, injecting particles into the trapping regions and replenishing the losses sustained by pitch-angle scattering. On a shorter time scale of days, storms and substorms introduce large time fluctuations in particle flux values and temporarily distort the equilibrium distributions expected on the basis of equation (9.1).

This simplified description of the impact of magnetic storms and substorms on trapped particles conceals a host of difficulties encountered in attempts at quantitative explanations. For example, the induction electric field, which must play a major role in particle acceleration, is difficult to model quantitatively. From Maxwell's equation, the electric field induced by a changing magnetic field is

$$\nabla \times \mathbf{E} = -\frac{\partial \mathbf{B}}{\partial t} \tag{9.2}$$

Even if $\partial \mathbf{B}/\partial t$ were known everywhere precisely, only the line integral of $\mathbf{E}$ about a closed path can be obtained from equation (9.2). In particular if $\mathbf{A}$ is the magnetic vector potential, then

$$\nabla \times \mathbf{E} = -\nabla \times \frac{\partial \mathbf{A}}{\partial t} \tag{9.3}$$

and

$$\mathbf{E} = -\frac{\partial \mathbf{A}}{\partial t} + \nabla \Psi \tag{9.4}$$

where Ψ is any scalar function of position. Establishing values for this scalar function requires additional assumptions about the electrical properties of the magnetosphere. In Chapter 8 the calculation of D_{LL} for magnetic fluctuations was accomplished by assuming that low-energy plasma would be redistributed along field lines so that $\mathbf{E}_\parallel = 0$ during the magnetic fluctuation. To cancel the induced $\mathbf{E}_\parallel$ the electrostatic potential field was arranged to give $\mathbf{E}_\parallel^{pot} = \partial \mathbf{A}/\partial t$. With the further assumption that the ionosphere connecting the feet of the geomagnetic field lines is a perfect conductor and therefore an equipotential, the electric potential Ψ at any point in space can be obtained by integrating $\mathbf{E}_\parallel^{pot}$ along a magnetic field line from the ionosphere to the point in question. Then knowing the electric potential Ψ everywhere, the total electric field can be obtained from equation (9.4).

In a major magnetic storm or substorm $\partial \mathbf{B}/\partial t$ may be qualitatively known, but the change in $\mathbf{B}$ may be so rapid that the low-energy plasma does not have time to alter its distribution and cancel the parallel electric field. Hence $\mathbf{E}_\parallel$ may not be zero and $\mathbf{E}_\perp$ cannot be derived from equation (9.4). This condition is most likely to occur at $L > 4$ where the thermal plasma density is low.

In addition to the induced electric field from $\partial \mathbf{B}/\partial t$ and the potential electric field derived from charges cancelling the induced field parallel to $\mathbf{B}$, other electric fields are important during storms. The solar wind flowing past the Earth imposes an electric field across the magnetosphere. In quiet magnetic conditions this electric field has little influence on the trapped particles except to distort their drift orbits slightly and to produce the radial diffusion described in Chapter 8. However, when this field is greatly intensified and varies with time it can also produce large changes in the L coordinate of a particle. If the change in L during a single drift orbit is small, cumulative changes produce radial diffusion. However, if the change in L is large in a single drift period, the overall effect will be observed as a sudden injection of particles.

Other processes probably occur in substorms but have been neglected in this simplified treatment. One such process is called magnetic reconnection. In locations where nearby magnetic fields point in different directions, for example near the neutral sheet in the geomagnetic tail, the magnetic fields may combine and annihilate, the energy of the magnetic field being converted to particle kinetic energy. This process is also believed to take place on the front surface of the magnetosphere when the solar wind magnetic field becomes connected to the Earth's field lines. In addition to accelerating particles, this process also enables solar wind plasma to penetrate the magnetopause and eventually enter the trapping region.

Another concept for particle acceleration is also dependent on strong gradients in the magnetic field near the neutral sheet. If the magnetic field changes direction or intensity over a distance smaller than the gyroradius, the first invariant will not be conserved and the gyromotion will be interrupted. Under this condition the particles may meander across the tail, moving between Northern and Southern lobes while being accelerated by the cross-tail electric field.

In major magnetic storms many additional factors may be important. The shock wave produced by the sudden compression of the sunward magnetopause propagates through the magnetosphere where it can interact with some of the trapped particles. The combined electric and magnetic fields present in a collisionless shock wave propagating through a plasma disrupt the adiabatic motion of charged particles and accelerate those whose drift phase is favorable. The shock magnetic field can also act as a moving mirror which both reflects particles and accelerates them. Finally, major magnetic storms result from a high-speed cloud of solar plasma impacting the magnetopause. These clouds are accompanied by energetic ions and electrons accelerated either in the solar flaring process or by the interplanetary shock wave at the front of the plasma cloud. In any case, during a magnetic storm the Earth is often engulfed by energetic particles, some of which may find their way into the magnetosphere and become trapped. Particularly during the initial stage of the storm, the magnetosphere is distorted and the configuration changes rapidly with time. The energetic solar plasma may thus behave in a non-adiabatic manner and become trapped deep in the magnetosphere.

In extreme cases, occurring perhaps once a decade, injection events associated with very large magnetic storms cause the sudden appearance of energetic ions and electrons at L values as low as $L = 2$–3. At these L values the injected particles remain trapped for months or years and are slowly distributed throughout the belts by radial diffusion. The relative importance of these spectacular, but rare, events in radiation belt formation remains uncertain, and the mechanisms causing the acceleration or injection are now being investigated.

At the present time a series of space experiments is in preparation to make comprehensive particle and field measurements simultaneously at a number of critical locations in the magnetosphere. Known as the International Solar Terrestrial Program, this mission will explore the changing nature of the entire magnetosphere and, in particular, the transfer of energy and particles from the solar wind through the magnetosphere and into the atmosphere. Spacecraft will be located in the solar wind to monitor the influence of the Sun, deep in the geomagnetic tail to observe particle and magnetic flux storage and release, and in positions to observe trapped particle populations and to measure the loss of particles into the Earth's atmosphere. Since the magnetosphere acts as a coherent system it is necessary to make measurements in all these regions simultaneously in order to fully understand the important processes.

The history to date of research on the Earth's radiation belts has been a series of unpredicted discoveries followed by interpretations and explanations which, after tentative acceptance, were frequently overturned when additional experimental data became available. Eventually the new results were accommodated into modified theories. Gradually, this process has led to a convergence of views on many aspects of the radiation belts. However, while the motion of charged particles in static electric and magnetic fields is fully understood, the richness of magnetospheric phenomena derives largely from the time-dependent electric and magnetic fields. As was discussed in Chapters 7 and 8, most existing theories for treating time-dependent effects are based on statistical properties of these fields and, furthermore, assume that the field fluctuations are small. Recent work has treated large field changes by numerical tracking of particles in idealized field geometries with specified time variations. This approach provides realistic answers for the cases considered and is leading to important insights on particle behavior in a realistic magnetosphere. Nevertheless, one should be prepared for further surprises, both in experimental and theoretical work. Future research will certainly alter some of the present concepts and may require the introduction of entirely new processes.

Appendix A

Summary of frequently used formulas

Numbers indicate where formula is located in text.

Motion of a charged particle in electric and magnetic fields

$$\mathbf{F} = q(\mathbf{v} \times \mathbf{B} + \mathbf{E}) \tag{2.1}$$

Gyroradius:

$$\rho = \frac{p_\perp}{Bq} = \frac{\gamma m_0 v_\perp}{Bq} \tag{2.4}$$

Gyrofrequency:

$$\Omega = \frac{qB}{\gamma m_0}$$

$$\Omega \text{ (radians s}^{-1}) = 1.758 \times 10^{11} \frac{B(T)}{\gamma} \qquad \text{(electrons)}$$

$$= 9.581 \times 10^{7} \frac{B(T)}{\gamma} \qquad \text{(protons)}$$

Guiding center drift velocities

Electric field drift:

$$\mathbf{V}_E = \frac{\mathbf{E} \times \mathbf{B}}{B^2} \tag{2.8}$$

Gradient B drift:

$$\mathbf{V}_G = \frac{m v_\perp^2}{2qB^3}(\mathbf{B} \times \nabla B) \tag{2.23}$$

Curvature B drift:

$$\mathbf{V}_c = \frac{m v_\parallel^2}{qR_c} \frac{\mathbf{n} \times \mathbf{B}}{B^2} \tag{2.25}$$

In regions where $\nabla \times \mathbf{B} = 0$ the $\mathbf{V}_c$ is given by

157

$$\mathbf{V}_c = \frac{mv_\parallel^2}{qB^3}(\mathbf{B} \times \nabla B) \tag{2.27}$$

Mirroring force:

$$\mathbf{F}_z = -\frac{mv_\perp^2}{2B}\frac{\partial B}{\partial z}\hat{\mathbf{e}}_z \tag{2.31}$$

$$= -\frac{\mu}{\gamma}\frac{\partial B}{\partial z}\hat{\mathbf{e}}_z \tag{4.20}$$

Dipole field equations ($B_0 = 3.12 \times 10^{-5}$ T)

$$B_r = -2B_0\left(\frac{R_E}{r}\right)^3\cos\theta \tag{3.13}$$

$$B_\theta = -B_0\left(\frac{R_E}{r}\right)^3\sin\theta \tag{3.14}$$

$$|B| = B_0\left(\frac{R_E}{r}\right)^3\sqrt{(1 + 3\cos^2\theta)} \tag{3.15}$$

Equation for dipole field line:

$$r = R_0\sin^2\theta \tag{3.17}$$
$$= R_0\cos^2\lambda$$

Particle motion in geomagnetic field

Bounce period:

$$\tau_b(s) = 0.117\left(\frac{R_0}{R_E}\right)\frac{1}{\beta}[1 - 0.4635(\sin\alpha_{eq})^{3/4}] \tag{4.28}$$

Drift period:

$$\tau_d(s) = C_d\left(\frac{R_E}{R_0}\right)\frac{1}{\gamma\beta^2}[1 - 0.333(\sin\alpha_{eq})^{0.62}] \tag{4.47}$$

where

$$C_d = 1.557 \times 10^4 \text{ for electrons}$$
$$= 8.481 \text{ for protons}$$

Adiabatic invariants

$$J_1 = \mu = \frac{p_\perp^2}{2m_0 B} \tag{4.13}$$

$$J_2 = J = \oint\mathbf{p}\cdot\mathrm{ds} \tag{4.31}$$

$$J_3 = \Phi = q\int_S\mathbf{B}\cdot\mathrm{dS} \tag{4.50}$$

Relationship of flux $j(\alpha, E)$ to phase space density $F(\mathbf{q}, \mathbf{p})$

$$F(\mathbf{q}, \mathbf{p}) = j(\alpha, E)/p^2$$

Electric and magnetic field transformations

Lorentz transformations of electric and magnetic fields are given for a primed system moving at velocity $\mathbf{V}$ with respect to an unprimed system. Parallel quantities are measured parallel to $\mathbf{V}$:

$$B'_{\|} = B_{\|} \qquad E'_{\|} = E_{\|}$$

$$\mathbf{B}'_{\perp} = \gamma\left(\mathbf{B}_{\perp} - \frac{\mathbf{V} \times \mathbf{E}}{c^2}\right)_{\perp} \qquad \mathbf{E}'_{\perp} = \gamma(\mathbf{E}_{\perp} + \mathbf{V} \times \mathbf{B})_{\perp}$$

where

$$\gamma = (1 - V^2/c^2)^{-1/2}$$

In magnetospheric applications $V < 10^5 \, \mathrm{m \, s^{-1}}$, $10^{-8} \, \mathrm{T} < B < 10^{-5} \, \mathrm{T}$, $E < 0.1 \, \mathrm{vm^{-1}}$. Therefore, to an excellent approximation $\gamma = 1$, $B \gg \mathbf{V} \times \mathbf{E}/c^2$ and it is permissible to use

$$B'_{\|} = B_{\|} \qquad E'_{\|} = E_{\|}$$

$$\mathbf{B}'_{\perp} = \mathbf{B}_{\perp} \qquad \mathbf{E}'_{\perp} = \mathbf{E}_{\perp} + (\mathbf{V} \times \mathbf{B})$$

Maxwell's equations

$$\nabla \cdot \mathbf{B} = 0 \qquad \nabla \cdot \mathbf{E} = \frac{\rho}{\varepsilon_0} - \frac{\nabla \cdot \boldsymbol{\rho}}{\varepsilon_0}$$

$$\nabla \times \mathbf{B} = \mu_0 \mathbf{i} + \mu_0 \varepsilon_0 \frac{\partial \mathbf{E}}{\partial t} \qquad \nabla \times \mathbf{E} = -\frac{\partial \mathbf{B}}{\partial t}$$

where $\mathbf{i}$ is the current density, ρ is charge density and $\boldsymbol{\rho}$ is the polarization of the medium.

Vector identities

$\mathbf{A}$, $\mathbf{B}$ and $\mathbf{C}$ are vector fields; $d\mathbf{S}$ is a surface element of S; $d\mathbf{l}$ is a line element of a contour around S; dV is an element of volume V:

$$(\mathbf{A} \times \mathbf{B}) \times \mathbf{C} = (\mathbf{A} \cdot \mathbf{C})\mathbf{B} - (\mathbf{B} \cdot \mathbf{C})\mathbf{A}$$

$$\nabla \times (\nabla \psi) = 0 \qquad \nabla \cdot (\nabla \times \mathbf{F}) = 0$$

$$\iint_S (\nabla \times \mathbf{A}) \cdot d\mathbf{S} = \oint_c \mathbf{A} \cdot d\mathbf{l}$$

$$\iiint_V (\nabla \cdot \mathbf{A}) \, dV = \iint_S \mathbf{A} \cdot d\mathbf{S}$$

$$\mathbf{A} \cdot \nabla \mathbf{B} = (\mathbf{A} \cdot \nabla) \, \mathbf{B} = \left(A_x \frac{\partial}{\partial x} + A_y \frac{\partial}{\partial y} + A_z \frac{\partial}{\partial z}\right)\mathbf{B}$$

In spherical polar coordinates

$$\nabla \cdot \psi = \hat{\mathbf{e}}_r \frac{\partial \psi}{\partial r} + \hat{\mathbf{e}}_\theta \frac{1}{r} \frac{\partial \psi}{\partial \theta} + \hat{\mathbf{e}}_\phi \frac{1}{r \sin \theta} \frac{\partial \psi}{\partial \phi}$$

$$\nabla^2 \psi = \frac{1}{r^2} \left[\frac{\partial}{\partial r} \left(r^2 \frac{\partial \psi}{\partial r} \right) + \frac{1}{\sin \theta} \frac{\partial}{\partial \theta} \left(\sin \theta \frac{\partial \psi}{\partial \theta} \right) + \frac{1}{\sin^2 \theta} \frac{\partial^2 \psi}{\partial \phi^2} \right]$$

Relativistic relations

$\mathbf{v}$ = velocity, Tm_0c^2 = kinetic energy, W = kinetic energy + rest mass energy

$$\beta = \frac{v}{c} \qquad \gamma = (1 - \beta^2)^{-1/2} = T + 1$$

$$m = m_0 \gamma \qquad \mathbf{p} = m\mathbf{v} = m_0 \gamma \mathbf{v}$$

$$W = (T + 1) \, m_0 c^2 = (m_0^2 c^4 + c^2 p^2)^{1/2}$$

$$\beta^2 = \frac{T(T + 2)}{(T + 1)^2}$$

Appendix B

Gyration, bounce and drift frequencies in a dipole field

The gyration frequencies ν_g and periods τ_g of low-energy electrons and protons in an idealized, symmetric dipole field are shown in Figure B.1 for altitudes up to 10^4 km. For relativistic particles these values should be divided by γ.

The bounce and drift periods of equatorial particles in a dipole field are given as a function of particle energy in Figure B.2 for electrons and in Figure B.3 for protons. Since the bounce period is inversely proportional to particle velocity (see equation (4.26)), the period becomes almost independent of energy as the electrons become relativistic.

For particles mirroring at any latitude, the bounce period can be expressed as

$$\tau_b(s) = 6.37 \times 10^6 Lg(\lambda_m)/v \tag{B.1}$$

where v is the particle velocity (meters s^{-1}) and $g(\lambda_m)$ is a geometrical quantity which depends on the mirroring latitude (or the equatorial pitch angle). A plot of $g(\lambda_m)$ is given in Figure B.4 as a function of both λ_m and α_E.

The azimuthal drift period for off-equatorial particles is expressed by the approximate formula (4.47). A graphical representation of the geometrical factor $\mathcal{D}(\lambda_m)$ as a function of mirroring latitude is given in Figure B.5. The drift period is

$$\tau_d(s) = \frac{\mathcal{D}(\lambda_m)}{L(E + m_0c^2)(v/c)^2}$$

where the kinetic energy of the particle E and the rest mass energy m_0c^2 are expressed in MeV. The ratio $(v/c)^2$ for any kinetic energy T (in rest mass units) is given by $(v/c)^2 = \beta^2 = T(T + 2)/(T + 1)^2$. For example, a 1 MeV electron has $T = 1.957$, $\beta^2 = 0.886$, and $E + m_0c^2 = 1.511$ MeV. If it has equatorial pitch angle of 90° and is at $L = 2$, $\tau_d = 5300/2(1.511)(0.885) = 1.981 \times 10^3$ s, in agreement with Figure B.2.

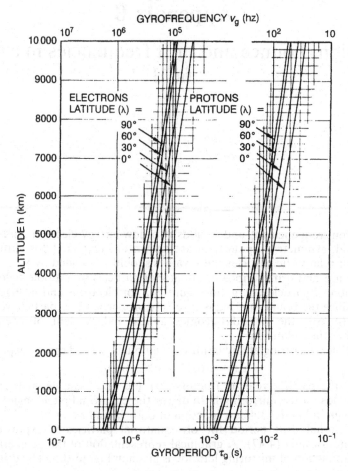

Figure B.1. Gyration frequencies and periods of trapped electrons and protons in a centered, dipole magnetic field.

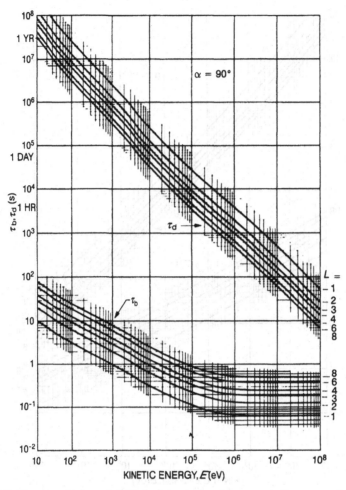

Figure B.2. Bounce (τ_b) and drift (τ_d) periods of equatorially trapped electrons.

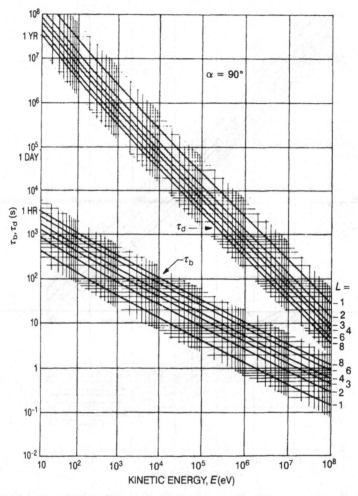

Figure B.3. Bounce (τ_b) and drift (τ_d) periods of equatorially trapped protons.

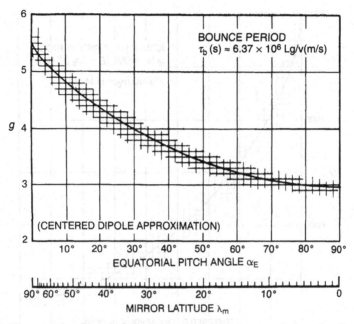

Figure B.4. Geometric factor for computing bounce periods of off-equatorial particles.

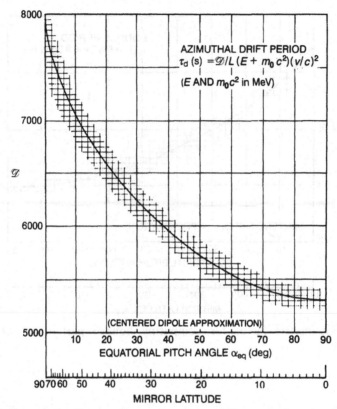

Figure B.5. Geometric factor for computing drift periods of off-equatorial particles.

References

Alfvén, H. and Fälthammar C-G. (1963). *Cosmical Electrodynamics*, Oxford University Press, London (Ch. 2).

Chandrasekar, S. (1943). Stochastic problems in physics and astronomy, *Revs. Modern. Phys.*, **15**, 1 (Ch. 6).

Fälthammar, C-G. (1965). Effects of time dependent electric fields on geomagnetically trapped radiation, *J. Geophys. Res.*, **70**, 2503 (Chs. 5, 8).

Haerendel, G. (1968). Diffusion theory of trapped particles and the observed proton distribution. In *Earth's Particles and Fields*, ed. B. M. McCormac, Reinhold, New York, p. 171 (Chs. 5, 8).

Lyons, L. R. and Williams, D. J. (1984). *Quantitative Aspects of Magnetospheric Physics*, D. Reidel, Dordrecht (Chs. 1, 2, 7, 8).

Northrop, T. G. (1963). *The Adiabatic Motion of Charged Particles*, Interscience Publishers, New York (Chs. 2, 4).

Roederer, J. G. (1970). *Dynamics of Geomagnetically Trapped Radiation*, Springer-Verlag, New York (Chs. 2, 4, 6).

Sawyer, D. M. and Vette, J. I. (1976). *AP-8 Trapped Proton Environment for Solar Maximum and Solar Minimum*, National Space Science Data Center, NSSDC/WDC-A-R&S 76-06, December (Ch. 5).

Schulz, M. (1993). The magnetosphere, Chapter 1 in *Geomagnetism* Vol 4, ed. J. A. Jacobs, Academic Press (Chs. 1, 3, 7, 8).

Schulz, M. and L. J. Lanzerotti (1974). *Particle Diffusion in the Radiation Belts*, Springer-Verlag, New York (Chs. 6, 7, 8).

Stern, D. P. (1976). Representation of magnetic fields in space, *Reviews of Geophysics and Space Physics*, **14**, 199 (Ch. 3).

Tsyganenko, N. A. (1990). Quantitative models of the magnetospheric magnetic field: Methods and results, *Space Science Rev.*, **54**, 75–186 (Ch. 3).

Vette, J. I. (1991). *The AE-8 Trapped Electron Model Environment*, National Space Science Data Center, NSSDC/WDC-A-R&S 91–24, November (Ch. 5).

Index

Printed in the United States
By Bookmasters